TECHNOLOGY AND ASSESSMENT
OF
SAFETY-CRITICAL SYSTEMS

Related titles:

Directions in Safety-critical Systems
Proceedings of the First Safety-critical Systems Symposium, Bristol 1993
Redmill and Anderson (eds.)
3–540–19817–2

SAFECOMP '93
Proceedings of the 12th International Conference on Computer Safety,
Reliability and Security, Poznań-Kiekrz, Poland, October 1993
Górski (ed.)
3–540–19838–5

TECHNOLOGY AND ASSESSMENT OF SAFETY-CRITICAL SYSTEMS

Proceedings of the Second Safety-critical Systems Symposium

Birmingham, UK
8–10 February 1994

Edited by
FELIX REDMILL and TOM ANDERSON

Safety-Critical Systems Club

Springer-Verlag
London   Berlin   Heidelberg   New York
Paris   Tokyo   Hong Kong
Barcelona   Budapest

Felix Redmill
Redmill Consultancy
22 Onslow Gardens
London N10 3JU, UK

Tom Anderson
Centre for Software Reliability
University of Newcastle-upon-Tyne
Newcastle-upon-Tyne NE1 7RU, UK

ISBN-13:978-3-540-19859-8   e-ISBN-13:978-1-4471-2082-7
DOI: 10.1007/978-1-4471-2082-7

British Library Cataloguing in Publication Data
A catalogue record for this book is available from the British Library

Typesetting: Camera ready by authors

34/3830-543210 Printed on acid-free paper

PREFACE

The programme for the Second Safety-critical Systems Symposium was planned to examine the various aspects of technology currently employed in the design of safety-critical systems, as well as to emphasise the importance of safety assessment and risk management in their design and operation.

There is an even balance of contributions from academia and industry. Thus, industry is given the opportunity to express its views of the safety-critical domain and at the same time offered a glimpse of the technologies which are currently under development and which, if successful, will be available in the medium-term future.

In the field of technology, a subject whose importance is increasingly being recognised is human factors, and there are papers on this from the University of Hertfordshire and Rolls-Royce. Increasingly, PLCs are being employed in safety-critical applications, and this domain is represented by contributions from Nuclear Electric and August Computers. Then there are papers on maintainability, Ada, reverse engineering, social issues, formal methods, and medical systems, all in the context of safety. And, of course, it is not possible to keep the 'new' technologies out of the safety-critical domain: there are papers on neural networks from the University of Exeter and knowledge-based systems from ERA Technology.

The crucially important field of safety assessment is represented by contributions from the University of Sheffield and Poland; there is a paper on safety management from Honeywell and Cambridge University; and these are balanced by papers on risk management and its economics from Alexander and Alexander, and on the safety case from Denmark.

Creating this programme, organising the symposium, and publishing the resulting proceedings has placed demands in a number of quarters. In particular, we would like to thank Joan Atkinson, Bob Malcolm, the authors of all the papers, and Linda Schofield at Springer-Verlag.

Felix Redmill and Tom Anderson
October 1993.

CONTENTS

The Safety-Critical Systems Club
sponsor and organiser
of the
Safety-critical Systems Symposium

The Safety-Critical Systems Club exists for the benefit of industrialists and researchers involved with computer systems in safety-critical applications. Sponsored by the Department of Trade and Industry and the Science and Engineering Research Council, its purpose is to increase awareness and facilitate communication in the safety-critical systems domain.

Responsibility for the organisation of the club was contracted to the British Computer Society and the Institution of Electrical Engineers, together with the Centre for Software Reliability at the University of Newcastle upon Tyne, and Felix Redmill, an independent consultant, is the Club's Co-ordinator.

At October 1993, there were over 1800 members, from both industry and academia, with 130 being from countries other than the U.K.

The Club's goals are to achieve, in the supply of safety-critical computer systems:

- Better informed application of safety engineering practices;
- Improved choice and application of computer-system technology;
- More appropriate research and better informed evaluation of research.

In pursuit of these goals, the Club seeks to involve both engineers and managers from all sectors of the safety-critical community and, thus, to facilitate the transfer of information, technology, and current and emerging practices and standards.

As well as initiating the Safety-critical Systems Symposium, the club runs at least four one- or two-day seminars each year, co-sponsors other events, publishes a newsletter three times annually, and provides ad hoc facilities of importance to the safety-critical systems community. In these ways it facilitates: the communication of experience among industrial users of technology, the exchange of state-of-the-art knowledge among researchers, the transfer of technology from researchers to industrial users, and feedback from users to researchers. It facilitates the union of industry and academia for collaborative projects, and provides the means for the publicity of those projects and for the reporting of their results.

To join or to enquire about the Club, please contact:
Mrs J Atkinson
Centre for Software Reliability
The University
20 Windsor Terrace
Newcastle upon Tyne, NE1 7RU
Tel: 091 212 4040; Fax: 091 222 7995.

A User's Perspective of Programmable Logic Controllers (PLCs) in Safety-related Applications

A.Greenway MSc DMS DipM AMIEE
Nuclear Electric plc
Gloucester, UK

Abstract

A personal and relatively informal view of the current and to some extent expected roles of PLCs in safety related systems is presented. An answer to the question : "Is there a role for PLCs in safety related systems?" is sought. The systems referred to here are plant or equipment control or protection systems. To answer this question, PLCs are explained for those who are not familiar with them and the term safety related is explored. The opportunity is taken to set out a basis for discussing safety related control and protection equipment by clearly defining terms such as reliability. The paper then sets out to answer the question proper by comparing two candidate solutions for a typical application. The application is described and the merits of the two candidates, one relay based and one PLC based, are compared. The impact that future developments might have are discussed.

1 Introduction

The Programmable Logic Controller (PLC) emerged in the 1970's as a marketing innovation which packaged microprocessor based components in a form for easy use in industrial control applications. Initially, functionality was limited to logic functions intended to replace relay circuits. Latterly, the functionality of the equipment has been greatly extended to provide for most of the needs of industrial control. PLC components are either used in bespoke plant control systems, typically mounted together with termination blocks, control switches and indication devices in boxes or cubicles and used for plant control or embedded within equipment manufacturers' products e.g. machine controls.

Almost from the day that PLCs emerged as distinct products, they have been employed by users within safety related systems.

The "User" referred to here is, for example, a project engineer charged with the responsibility for providing plant operators, with suitable control equipment such that statutory and economic criteria are met. Typical statutory criteria arise from pollution controls or the Health and Safety at Work Act. Economic criteria include capital costs (equipment and engineering), operating costs and operating performance.

As time has gone by, the safety environment within which these equipments have been used has been changing as society and individuals have changed their understanding of safety as it impinges on their applications and this technology. These changes have been most pronounced in the area of software engineering as the engineering community has attempted to come to grips with concepts such as software reliability and formal methods.

This changing environment has had a profound impact on the ability of users to predict the costs and timescales of development or refurbishment projects using microprocessor based equipment in a safety related role. This is exacerbated by the nature of such projects. Whereas in a non-safety related role, engineering costs would be comparable to equipment costs, in a safety related role engineering costs dominate equipment costs and it is not uncommon for the safety aspects of a project to dominate all costs. Such has been the rate of change, there has been a real concern that there might not be a role for microprocessor based equipment in such systems at all.

However, much of the work that has been undertaken in the last few years is now beginning to reach the stage when it is believed that there is some real understanding of the technology. Standards are emerging and the rate of change of ideas is slowing. It therefore seemed appropriate to review the current position with respect to one of the microprocessor based solutions to control problems currently available i.e. the PLC. The reason that PLCs were chosen is that, potentially, this approach offers a very cost effective solutions to many problems.

The objectives of this paper are to set out a personal and rather informal view of the current and to some extent expected roles of PLCs in safety related systems when compared with other electrical/electronic technologies.

This paper seeks to answer the question : "Is there a role for PLCs in safety related systems?". Here, the systems referred to are plant or equipment control or protection systems. To answer this question, PLCs are explained for those who are not familiar with them and the term safety related is explored. The opportunity is taken to set out a basis for discussing safety related control and protection equipment by clearly defining terms such as reliability. The paper then sets out to answer the question proper by comparing two candidate solutions for a typical application. The application is described and the merits of the two candidates, one relay based and one PLC based are compared. The impact that future developments might have are discussed.

2 What is a PLC?

A Programmable Logic Controller is a packaged set of microprocessor based control components comprising analogue and digital input modules of various types, digital output modules of varying capabilities, power supply conditioning and processing modules, together with design and testing tools.

Input and output modules are intended to be user configured, providing varying degrees of signal conditioning and fault monitoring and they are characterised by standard industrial interfaces and robustness e.g. an ability to withstand common and series mode electrical noise.

The processors and supporting equipment are designed to meet industrial needs where unattended operation is the norm. This implies high reliability, an ability to detect and signal failures (to initiate repair) and failure modes which will not hazard plant or equipment.

The on-line features are typically provided by executive firmware which provides a user friendly interface to I/O, real time facilities, support for communications and an interpreter for the application program.

The equipment is configured using programming languages which are typically interpreted in real time. Simpler PLCs execute the code sequentially without any multitasking. More complex PLCs provide facilities for more sophisticated manipulation of control flow. There are a number of languages available, and for many years these have been the subject of marketing pressures such that until recently, no standards had emerged. IEC Committee 65B, WG 7 has recently published a standard for Programmable Logic Controller languages, IEC 1131, Part 3 [1], and it is expected to have a significant impact on the design of software for PLCs. Although it does not constitute a great departure from current practice, it could potentially provide the impetus for the development of more sophisticated tools with associated improvements in software quality, complexity and programmer/designer productivity. It will have an impact upon practicing systems designers, PLC programmers, tool suppliers, equipment suppliers and educationalists. Issues likely to be raised are the language standard itself, expected costs and benefits, compliance issues, suitability issues (e.g. for machine control, process control, batch control) and performance issues.

IEC 1131-3 defines five PLC programming languages, three of them with a graphical definition and two being text based:

-       Instruction List (IL)

        A low level assembler type of language (textual, without support for structuring)

-       Ladder Diagram (LD)

        This is a variant on the well established languages based upon relay drawing and design conventions (graphical, with a very limited support for structuring)

-       Function Block Diagrams (FBD)

This defines a program in terms of control blocks which process input variables to produce one or more outputs. Users can build up libraries of such blocks (graphical with some limited support for hierarchical structures)

- Structured Text (ST)

    A high level language that can be used to create application specific function blocks (textual, structured)

- Sequence Function Charts (SFC)

    A high level graphical language intended to provide a vehicle for manipulation of threads of control and which is suited to sequence and multithreaded programming (graphical, can be used to structure code)

This list shows a progression from low level unstructured languages towards higher level languages, offering greater opportunities for imposing a well defined structure on code.

The extent to which the standard has been adopted is not very clear, but a number of manufacturers have products conforming to one or more of the language standards and others have indicated an intent to develop products that will.

PLCs are poorly supported by software tools when compared to other software approaches. In particular, tool support for testing is weak e.g. test harnesses and test coverage tools are not available. Tools are available to aid documentation and principally to cross reference use of variables. The most recent developments have occurred in the areas of simulation and emulation with a view to improving testing.

3 Safety and Reliability

The opportunity is taken here to explore some of the concepts used in this paper with a view to very clearly setting a foundation for later discussion.

The term 'safety related' has been coined to cover systems which if they fail might result in a hazard of limited proportions such that unreliabilities of the order of 10^{-1} - 10^{-2} per annum or 10^{-2} - 10^{-3} per demand are broadly acceptable (these measures are discussed later).

Some basic requirements for safety related control or protection systems are:

- The system should meet some functional fitness for purpose criteria;

- The measure of expectation that the equipment will fail to meet these criteria during a specified period of time or when operation is demanded should be small;

4

-      The system should meet the above requirements in the face of expected hazards;

-      It should be possible to reasonably demonstrate that the system satisfies the above requirements even taking into account limitations of any assumptions made.

The second of these requirements concerns reliability. There are two definitions of reliability which are germane:

-      A measure of expectation of failure on demand. This definition is most appropriate when there are intermittent requirements to operate which are discernable in time e.g. demands on a protection system.

-      A measure of expectation of failure within a period. This definition is most appropriate when the requirement to operate is continuous. (Note that this does not to assume a constant failure rate)

It is well recognised that the causes of failure can be broadly categorised into two types: design faults and equipment degradation. The former cause is characterised by it persistent existence (until corrected that is). The latter cause is characterised by its transient nature e.g. random hardware failure. Some design faults lead to changes in failure characteristic and so involve elements of both of these. Surprisingly, the transient faults have proved to be the easiest to handle. This has arisen because of the simplicity and effectiveness of the constant failure rate model for hardware. However, what is missing is a suitable model for design faults. One is proposed as follows:

The concept of operating space is introduced. This is the space defined by the inputs/states/outputs of the plant and system under consideration. It includes all states that might arise during the lifetime of the plant and system. This is illustrated in Figure 1 which is intended to represent a simplified system having one input and one state variable. Any particular input/state/output combination is represented by a point in the operating space. This concept is frequently encountered in control system engineering and has been developed in the context of software systems in [2].

The design space is an area within the operating space within which the plant or system could be expected to operate. It will encompass variation, for example, in inputs or environment. An assumed design space is that used as a basis for the equipment design. This corresponds to the area where designers intend the plant to meet the fitness for purpose criteria. It may not correspond to the real design space. For example, there may be a real requirement for equipment to operate to 75 degC when the design has assumed that 70 degC would be adequate.

Plant also has a characteristic that from one moment to the next, the operating points are typically correlated (although this is not necessarily so). This allows the concept

5

of an operating trajectory to be introduced. This is a trajectory through the input/state/output space traversed by the system in time.

Typical systems have the characteristic that there will be a greater probability of finding the plant in some areas of this space than others. For example, the area outside of the design space is where the plant and equipment is not expected to operate. The expected activity might be represented by the use of activity contours which indicate areas where the trajectory would be expected to be found.

It is accepted that the system contains design errors. These are areas within the operating space where the system does not satisfy the fitness for purpose criteria even though this was intended. These errors arise from two principal causes:

-       There is an error within the system design (typically revealed by verification);

-       There is an error in the understanding about the fitness for purpose criteria or the boundary of the design space (typically revealed validation).

Failures occur when the operating trajectory encounters a design error. Errors are considered to occupy volume in the operational space. These are referred to as 'error crystals' in [3].

It is also observed that there are areas of the operating space which are frequently visited and areas which are infrequently visited. It is clear that errors in areas of the operating space which are infrequently visited would be expected to result in correspondingly low failure rates.

When there is a continuous requirement to operate, a failure will occur as soon as an error is encountered. It is seen that the time period reliability relates to the speed that the trajectories traverse the operating space and the preponderance and distribution of errors in the design space.

The case when there is an intermittent requirement to operate is different. Demands will be correlated with particular areas within the operating space. It may be that errors coexist in these areas. Here, trajectories might encounter demands where there are no errors (no failure), errors where there are no demands (no failure) or error and demand (failure). Reliability is related to the preponderance of errors and their correlation with demands. This measure of reliability has no inherent connection with the speed with which the trajectory is traversed.

Whilst the complexity of most systems prevents the development of the model very far, it serves as a foundation upon which discussion of such issues as inspection and testing and the role of in-service experience can rest. For example, it aids visualisation when discussing effectiveness of verification, testing, inspection, validation and in-service experience for say the application software and the

6

executive firmware.

The model is also limited in that it does not seek to represent the consequences of an encounter between the trajectory and an error.

In the context of projects involving PLCs in safety related applications, the design faults of most concern are errors in the application software and executive firmware. It is in these areas that safety issues tend to have most impact on project management.

Given that this paper compares a programmable with a non-programmable approach, there is some benefit in separating the system design issues (common to both) from the software design issues (only relevant to the PLC). Although many lifecycle models use the words system design and software design, often their definition is vague. Here there are quite definite understandings of the terms as related to processes (essentially decision making processes) and the results of those processes.

It is proposed that the system design is a model of the intended functionality, performance and reliability of the system and/or set of constraints that the system must meet. The system design process is a process of establishing this model/set of constraints based upon general fitness for purpose criteria, application specific criteria, operating criteria, maintenance criteria etc. It is proposed that the software is defined as the set of instructions or tokens or similar actually interpreted by the processor. The software design process is a process of converting the system design model into software such that the delivered system will perform in the same way as the model and/or satisfy the design constraints.

4 Case Study - Fuel store

It is rather difficult to consider the role of PLCs entirely in the abstract. Therefore, an example of a plant which could potentially employ either PLCs or relays in a safety related role is given.

The case offered is concerned with a facility to store reactor fuel elements pending further processing. Its safe operation requires that fuel elements are cooled. The primary coolant is pressurised CO_2. This is provided to the facility from a redundant supply via some 13 valves. The valves are used for filling various parts of the facility, controlling pressures and blowing down.

Heat is removed from the storage facility by a secondary cooling system using a demineralised water jacket forming part of a duplicated re-circulating cooling system. The heat is then dumped via four heat exchangers to a tertiary sea water system. The secondary system comprises two sets of main/standby circulating pumps, make-up pumps, valves (26), tanks and heat exchangers (4). The tertiary cooling system also comprises two sets of main/standby circulating pumps, valves (14) and heat exchangers (4).

7

Typical control requirements would be the control of CO_2 pressures or control of secondary cooling system feed tanks within defined limits. Typical protection requirements would be initiation of pump shutdown on fault detection, starting standby pumps or monitoring of tank levels to raise low level alarms to initiate remedial action.

A typical PLC response to these requirements would involve two PLCs providing control functions on a main/standby basis and a further two PLCs providing protective functions on a 1 out of 2 basis (i.e. operation of either provides a safe action). Each PLC would incorporate a ladder diagram program comprising some 250 rungs with up to 12 contacts on each rung. There might be 400 digital and analogue inputs and some 80 digital outputs. Safety functions would be provided by most of the software, while perhaps 10% of the software would be dedicated to maintenance operations when no direct hazard could occur in any case. The PLCs would incorporate features to display alarms and indications and perform self test functions.

An alternative solution might involve two sets of relays (300 in each) and analogue trip amplifiers (40), performing the same tasks as the PLCs above, having some limited capability for fault detection.

5 Evaluation of PLC versus relay solution

Each of the candidate solutions has been evaluated. In neither case have sensors, cables, power supplies, actuators etc. been explicitly considered as these are common to both approaches.

The evaluation comprised three basic elements:

-       Hardware reliability

        The example systems were evaluated with the following results:

        PLC: typically 2.5 failures per annum - 97% failures revealed - 50% unrevealed failures safe - dangerous failure rate of $3.75*10^{-2}$ per annum

        Relay: typically 1.5 failures per annum - 50% failures revealed - 90% unrevealed failures safe - dangerous failure rate of $7.5*10^{-2}$ per annum.

        In both cases, significant numbers of relay contacts are employed and it is judged that the lifetimes of these would be more constraining in the case of the relay solution, given the lesser degree of monitoring.

        Although the failure rates are different, given the broad nature of the judgements made, this is not believed to be significant.

        The use of a PLC in a smaller application could see hardware reliabilities of 0.2 failures per annum - 97% revealed - 50% unrevealed failures safe - dangerous failure rate of $3.0*10^{-3}$ per annum (or $3.0*10^{-3}$ per demand for protection given a 2 year proof test interval).

Vulnerability to hazards

In general one would expect similar degrees of invulnerability to most hazards from either solution. The one significant exception is in the area of electromagnetic interference. The PLC system is potentially the most susceptible approach. However, adequately resistant equipment can be purchased, which when employed in suitable enclosures will continue to function even when subjected to this hazard. In either case, the typical response to hazards is to design and qualify (i.e. test) against them.

Susceptibility to design error

This is addressed on the basis of a design aim that design errors should not dominate system unreliability.

Referring back to section 3, it is believed that the system design processes and resulting system design would be similar whichever implementation was chosen. There is also an argument that the relay hardware design would be very similar to the PLC software design. However, while the possibility of "sneak circuits", logic error and wiring error is clearly a possibility in the relay solution, there is not the potential for faults comparable to corruption of data by inadvertent use, timing problems, control flow problems or underlying firmware problems. On the assumption that the relay based solution will not suffer greatly from design error, it remains to evaluate the PLC solution in this respect.

The candidate strategies for addressing PLC design error are as follows:

(i)   Assess the characteristics of the PLC design and its development process with a view to inferring reliability and assess the characteristics of the design and its maintenance process with a view to extrapolating this result into the future. This is the approach currently advocated by IEC Committee 65A.

(ii)   Assess the in-service experience of the design with a view to inferring future reliability.

The strategies are not mutually exclusive and in practice, both would be expected to make a contribution to a judgement about the expected reliability.

9

For PLC executive firmware, although current practices for the maintenance of production standard firmware have often been found to be reasonable, it is usual for the designs to have undergone considerable development over a long period of time, with an initial point being prototype software. The nature of PLC's i.e. produced and used in large quantities, suggests that a large amount of in-service experience should be available. PLC hardware components are typically designed using well established techniques. Even though it would be possible to make claims on this, the availability of in-service experience provides an alternative basis for judging the quality of the equipment. Therefore the strategy selected for executive firmware and hardware is one which depends on in-service experience. This is pursued in Appendix A, with the conclusion that it would be reasonable to make claims for PLC executive firmware and hardware in safety related applications.

Although a decision to use a PLC is made early in the process, this could reasonably take account of later application testing which would provide further evidence about the absence or otherwise of errors within the design.

In contrast to the executive firmware and hardware, typical characteristics of overall system design and application software design include an opportunity to influence the design processes and relatively little in-service experience. Therefore, the strategy selected for system and application software design depends upon judgements about the design processes. The main objectives set for such processes are as follows :

-       The completeness and accuracy of the application requirement specification (including functional, performance, reliability, tolerance of hazards clauses);

-       The adequacy of the design process:

    *   The capability of the design process to capture the application requirements;

    *   The capability of the design process to convert the applications requirements into a system design and hardware and software requirements;

    *   The capability of the software design process to convert the application software requirements into an implementation representation;

    *   The capability of the hardware selection and configuration process to convert the hardware requirements into a design;

* The capability of the production process to integrate the implementation with the hardware and executive firmware.

- The adequacy of the safety assurance process

   * The capability of the verification process to detect design errors;

   * The capability of the validation process to detect requirements specification errors;

   * The provision of assurance.

- The maintenance of system integrity during operational lifetime

Each of these issues is affected by a design decision to use a PLC. The evaluation of the extent to which PLCs might be expected to satisfy these objectives is pursued in Appendix B. Here a detailed assessment of a typical PLC development process is reported. This concludes that it would be reasonable to claim that available systems and application software design techniques were suitable for safety related applications. The limiting factors are perhaps the lack of an integrated development environment extending to formal methods and the poor support for testing. The provisos are less significant when PLCs are used in simple applications e.g. the minimal size application noted above.

6 Future developments

Future developments are expected in the areas of formal methods. [4] reports the development of an interactive system with a graphical interface for constructing and validating PLC application software written in the Function Block and Structured Text languages. The semantics of a graphical design is defined by a mapping associating each design element with a formal specification. The specification provides the basis for rigorous proofs and early tests of critical properties of a new design. A research prototype was the basis of the reference, but it gives an indication of products which may be available in the future. The availability of such products would strengthen the role that PLCs could adopt in safety related systems.

A significant shortcoming of PLCs which has emerged is the lack of an integrated approach to design and testing. It is hoped that the publication of IEC 1131-3, will provide the catalyst to such developments.

7 Conclusion

This paper sought to answer the question : "Is there a role for PLCs in safety related systems?". The opportunity has also been taken to set out a basis for discussing safety related equipment by clearly defining terms. The two candidate solutions for a typical application have been compared in terms of:

1. Expected hardware reliability

2. Vulnerability to hazards

3. Design errors

with the conclusion that it would be reasonable to consider either a programmable or non-programmable solution.

The evaluation also points the way towards a systematic approach for the management of engineering projects employing PLCs in safety related roles. The type of assessments carried out here could be proceduralised and carried out for any project.

The impact that future developments might have has also been discussed.

Overall, it would seem that it is quite feasible to claim that PLCs are potentially suitable candidates against which to make modest safety claims and there is some expectation that this will improve in the future.

References

1. IEC 1131 Programmable Controllers - Part 3: Programming Languages. First edition. 1993.

2. Bishop P.G, Pullen F.D. Software Test and Evaluation Methods Project (STEM) - A Study of Software Failure Behaviour. National Power Research Report unrestricted. ESTD/L/0074/R89

3. Finelli B.F. NASA Software Failure Characterization Experiments. In: Reliability Engineering and System Safety, Elsevier 32 (1991) pp 155-169

4. Halang A.H, Kramer B. Achieving High Integrity of Process Control Software by Graphical Design and Formal Verification. In: Software Engineering Journal. IEE January 1992 pp 53-64

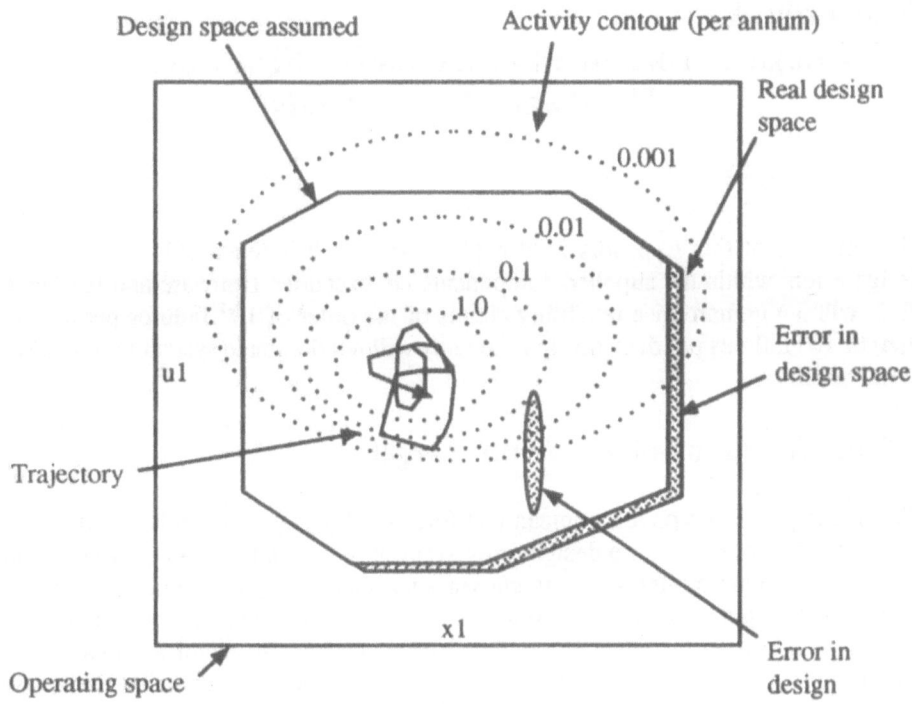

Figure 1   **Operating Space**

Appendix A
Design Errors in PLC Executive Firmware and Hardware Components

A.1 Introduction

The objective of this Appendix is to explore whether it is reasonable to claim that design errors within the supplied components i.e. executive firmware and hardware PLC, will not compromise reliability claims of the order of 10^{-2} failures per annum (fpa) or 10^{-3} failures per demand (fpd) i.e. they will not dominate system unreliability at this level.

A.2 Interpretation of Reliability Targets

The first target is interpreted to mean that for, say 500 years accumulated operation of a PLC, it is tolerable for a design error to result in one failure. The period of 500 years is somewhat arbitrary, but is chosen such that this type of failure could be claimed not to dominate system unreliability. It is therefore claimed that 5000 years accumulated experience should provide a sufficient basis to claim an appropriate level of reliability. No specific justification for this is provided except that the period for which experience is required is some 10 times the target MTBF. This factor is judged to be appropriate to give some confidence in the claim (if the failures occurred at random intervals, such a period would lead to better than 85% confidence) and to make some allowance for the lack of maturity of software reliability modelling.

The second reliability target is interpreted to mean that for say every 5000 demands made upon the PLC, it is tolerable for a design error to result in one failure. As above, the number of demands is somewhat arbitrary. It is therefore claimed that 50000 demands accumulated experience should provide a sufficient basis to claim an appropriate level of reliability. The period for which experience is required is some 10 times that target again as above. In this context i.e. a safety related application, a demand is a set of circumstances which require a certain specified operation from the software, which if not provided will result in a hazard. The nature of the specification implies that the demands are infrequent (in time).

It should be noted that no claim has been made for particular failure mode frequencies. In practice, it would be expected that failures in a hazardous sense would only comprise a fraction of the total. However, even with an understanding of the application it would not be possible to make any significant judgement here.

Two issues connected with in-service experience remain to be resolved; how is in-service experience to be measured in general and in particular how is it to be measured such that it can be related to the 10^{-3} fpd target.

For in-service experience to contribute towards a judgement about reliability, it must be relevant and there should be an effective mechanism for the recognition, identification, feedback to manufacturers and documentation of faults. With regard to relevance, it is generally accepted that experience of a population of similar items can contribute to judgements about the behaviour expected of individual equipments. This potentially allows the in-service experience of the entire population of the PLC type to be taken into account. However, this implies that relevance of the in-service experience must also be taken into account. This can be judged by comparing relevant features of the application under consideration with those features of the equipment in the field that are being claimed in support of the in-service claim. Of course, it is likely to be very difficult to obtain information about the equipment in the field and therefore, some judgement about the expected use of the features of the application under consideration is required.

The issue of errors identified but corrected must be addressed. It is considered that providing the number of errors is small, some degree of reliability growth can be claimed on the assumption that once corrected, the design is as if the error did not ever exist. This pre-supposes that no new errors are introduced and it is for this reason that the small number of errors is referred to. Given that a reasonable modification process is in place e.g. one consistent with ISO 9001, it is claimed that the entire in-service experience could reasonably contribute to the judgement about reliability as if there had been no errors.

A particular issue concerns the interpretation of in-service experience in terms that may be related to the failure per demand target. Different approaches are proposed for executive firmware and hardware.

It is proposed that for PLC executive firmware, there is nothing to distinguish a demand related to safety from any other demand. The justification for the approach in general is that the executive firmware is contained within a relatively protected environment such that any demand for its operation is much like any other. Demands arising from hazards are not expected to particularly 'stress' the firmware (cf fire detection) and would appear like any normal operating demand. It is therefore argued that operation of a PLC, for example in a machine tool application, provides evidence that the firmware is capable of responding to a large number of demands successfully and further, that operating experience of that part of the firmware associated with input-processing-output activities accrues with every cycle of the PLC (which typically lasts 200ms), although for some other roles, the experience accrues more slowly. Software reliability has been related to the extent to which the PLC moves around its input and state space i.e. simply repeating the same set of operations provides very little evidence of general reliability. For machine tool applications, the PLC would be expected to traverse a significant part of its input and state space in periods ranging from 2 minutes to 20 minutes. It is proposed that this concept be used as a basis for making judgements about the extent of in-service

15

experience necessary. It is therefore proposed that it is not unreasonable to claim that for most applications, the input and internal state will vary significantly at least once per day. The target for accumulated operating experience set out in section A.2 is 50000 demands. At the rate of one per day, this would be equivalent to 137 years accumulated experience. For some features of the firmware e.g. connection to programming tools, this claim is not realistic and it is therefore necessary to make a judgement about the relative frequencies of the use of the feature in the application and the in-service experience.

Most of the above applies equally to hardware components, however, it would not be true to say that they were impervious to environmental stress. This is covered separately in the section on vulnerability to hazards.

From the above, it is clear that to make a case based upon in-service experience, some judgements must be made both about the application with a view to determining whether any special features or operating regime are employed and about the likelihood that failures would be expected to be observed, identified and reported. This involves a failure modes and effects analysis (FMEA) to support limited arguments that failures would be expected to be revealed and/or that failures would not be expected to result in a dangerous failure mode.

A.3 Evaluation of PLCs

One of the strengths of PLC equipment is that it is produced in large quantities. Production runs of 5000 at rates of 1000 per annum are not uncommon. Taking this factor alone into account, some 3-4 years production should provide a sufficient basis upon which to make the claims required. Further, new products are often derivatives of older products and it would seem reasonable to take account of this factor also.

There are, however, some general points that arise from the FMEA type of exercise referred to above which should be taken into account :

-       Some PLCs allow the use of a programming tool for testing and debugging. Such use would be expected to be infrequent and associated faults would be expected to be attributed to human error. Faults would therefore be poorly represented in the in-service record. The response to this would be to make an appropriate allowance in the extent of in-service experience required, to make arguments that the frequency of intended use is low or to claim suitable preventative controls.

-       Infrequent transient faults would be expected to be attributed to hardware. Again such faults would be expected to be poorly represented in the in-service record. The response to this would be to make an appropriate allowance in the extent of in-service experience required or to place constraints on the software design such that, for example, internal storage of state is avoided.

Appendix B
Evaluation of PLC System and Application Software Design Processes

Each of the objectives for system and application software design identified in the main body has been applied in turn to a typical PLC, with the following results:

Adequate Specification of Safety Requirements

This might be considered to be outside the scope of this assessment, however, there are issues arising from a decision to use a PLC which could affect requirements specification. For example, for the highest reliabilities, formal methods suitable for the design process are desirable. There are no formal design methods that are generally applicable and commercially available that are suitable for direct use with the PLC languages. It can be argued however, that for simple logic systems, there is a propositional logic that will support application level ladder diagrams. On the positive side, for many PLCs, there are simulation/emulation packages which allow the animation of application functional requirements. The limits on the practical extent to which specification of requirements can be made only prevent a PLC system being claimed at the highest levels of reliability.

Adequacy of the Design Process

Requirements capture

As noted above, lack of development in the use of formal methods limits the claims which might be made.

System design and software requirements documentation

A number of system design issues are relevant:

-        Fault detection and tolerance

         PLCs can be used in a redundant fashion to provide fault tolerance. This fault tolerance can be achieved either by application configuration or by employing PLCs which themselves embody redundant architectures. Many PLCs employ built in self test which is either activated at start-up/reset or periodically during normal operation. Communication links typically include error detection mechanisms and application level checks can be implemented if necessary. A significant issue concerns the coverage of self tests. Experience has shown that significant pessimisms have to be assumed because suppliers find themselves unable to be specific about self test

17

coverage. Many software checks can be built into application level programs, however, such an approach tends to compromise system complexity.

- Design for security and maintainability

Many issues here can be addressed at the system design level (at a cost); application level checks can confirm correct and complete configurations; access controls to cubicles can limit the possibility of error. One of the most limiting features of PLCs is typically the lack of inherent protection from and consequences of errors in inserting replacement components.

- Design for operability

All issues here are in the domain of the application designer and there are few characteristics of PLCs which would have a bearing. However, the performance of many of the operator interfaces designed for use with PLCs are somewhat limited in terms of timeliness of update and consistency with in-house requirements.

- Software requirements documentation

Again as noted above, lack of development in the use of formal methods limits the claims which might be made. Typically, software requirements are documented using structured English. There are no overriding barriers to more rigorous approaches other than the lack of trained staff.

Software Design

A number of software design issues are relevant:

- Design for Simplicity and Predictability

Insofar as the facilities provided for the application designer by the PLC itself is concerned, it is not possible to generalise. Some of the smaller PLCs only require simple operation. Many PLCs do not support multiple threads of control and most often only employ a single thread of control through the interpreter, using interrupts to handle timing functions and communication functions. Most PLCs employ a simple input-process-output model which is quite predictable.

As noted in the main body of the paper, there are a number of languages available. The most popular language, ladder logic, severely limits the extent to which structuring of the software design is enforced. (This does not prevent some measure of structuring and rules for ladder logic design can be imposed). The most significant handicap is the global nature of data and

18

the lack of protective mechanisms intended to limit the effects of errors. (This is limited to some extent because many PLCs have available cross referencing tools).

- Clear Design Description

Lack of support for structured i.e. hierarchical and modular, documentation is the most limiting factor.

- Verifiability of design

With the publication of IEC 1131-3, there is now a reasonably complete and unambiguously defined language which can form the basis of verification. Its standardisation offers the expectation that independent verification might be easily procured. Limitations in the structuring of the software potentially limits the claims that might be made.

- Tools

In contrast to other environments, there are no tools available for automatic code generation from high level control system abstractions.

Hardware selection and configuration

In general, PLCs are available with comprehensive specifications and instructions for configuration. Few PLCs have been subject to the degree of independent test desirable for a safety related application, although it is quite straightforward to procure such tests.

Integration with the hardware and executive firmware

In general, application programs will be loaded from a development/documentation tool over a serial link into the PLC battery backed RAM storage. For safety related applications, EPROM may be preferable. In either case, configuration management support can be provided.

The adequacy of the safety assurance process

Verification

For the application software, it is feasible to make the most formal proofs provided the design of the software is suitably simple. The extent to which this can be carried out is limited by the availability of tools to accomplish this task. None of the IEC languages have any established safe subset, although it is believed to be relatively easy to address ladder diagrams in this respect. Control flow, data flow and information flow analysis is feasible although again this suffers from lack of tool

19

support and the lack of structuring supported by some languages. The imposition of coding standards is feasible, with the caveat that there are no established bases for setting these out. Testing is clearly feasible, but metrics in the form of path coverage etc., are not well established, nor is testing well supported by tools. None of these limitations actually prevent verification of small applications. The respective roles of inspection and test are particularly unclear for ladder diagrams.

Validation

In general, it is claimed that validation is implementation independent. This is in part supportable providing the complexity admitted by a PLC solution is not exploited. In any case, emulation and simulation tools do exist to support this activity. There are, however, some particular issues that need to be addressed here. These concern the failure modes of PLC systems e.g. consequences of electromagnetic interference, and the impact on the operation of the software. It is unlikely that extensive fault injection studies could be undertaken, although it is feasible to undertake limited exploration of the consequences of changes in the environment. While failure of PLC hardware is covered by alternative arguments i.e. in-service experience, this does not address the impact of those failures on the application. Provided extreme claims for self test coverage are not made it is believed that this is reasonable.

Safety Assurance

The main issue to be addressed here concerns the extent to which a history of reliable operation could be adopted as part of the safety argument. This is addressed in Appendix A.

Operation and Maintenance

There are a few issues to be addressed that are special here. Clearly, there are limitations in the media used to store programs and there are features provided for debugging e.g. the forcing of inputs and outputs that must be addressed in administrative controls. Some PLCs are better here than others; they remove all forcing when a key is switched to normal operation.

It is concluded that it would be reasonable to claim that available systems and application software design techniques were suitable for safety related applications. The limiting factors are perhaps the lack of an integrated development environment extending to formal methods and the poor support for testing. The provisos are less significant when PLCs are used in simple applications.

Methods and Techniques of Improving the Safety Classification of Programmable Logic Controller Safety Systems

Author
C J Goring
August Systems Limited (UK)

Abstract

This paper provides guidelines on the approaches that can be taken to improve the safety classification of Programmable Logic Controller Safety Systems.

For reference the VDE 801 Classification, the draft IEC 65A Working Group papers and the ISA SP84 draft documents are taken as providing the safety metrics to be achieved.

1.   Introduction

For the information in this paper to be placed in context we must first review the classification references produced by DIN VDE 801, IEC 65A and ISA SP84, the following sections of the paper will then refer to these classification specifications and provide information for the more subjective criteria.

Having reviewed the classification criteria it will then be practical to proceed to provide by example some configurations of Programmable Logic Controllers with associated components to meet these classification requirements. Additionally a few of the configuration anomalies are highlighted and a view is placed upon the relevance of these areas to safety.

In conclusion the importance of unified standards is reviewed and the author looks forward to final anticipated international standard being widely accepted.

2.   System Classification Analysis

There are a number of classification standards that have been and are being developed. Many of these standards are industry specific, such as those national standard associated with the Nuclear Industry. For the purpose of this analysis we will only consider the two non-industry specific classification standard, DIN VDE 801, IEC 65A (Draft) and the process industry standard ISA SP84 (Draft).

Both the IEC 65A Draft (WG10) and ISA SP84 (Draft are converging upon the same definition of integrity levels. Each standard defines four (4) levels of safety integrity. Unfortunately in the definition of classification for integrity class selection a certain amount of subjective judgement has to be used. An example would be the definition of say a minor injury, and terms such as "almost impossible".

The DIN VDE 801 specification provides eight (8) levels of safety classification, there is however some correlation between pairs of classification levels in the DIN VDE 801 specification and the IEC 65A WG10 document.

For the purpose of clarity for the remainder of this paper, I will refer only to the IEC 65A WG10 document and in particular Annex A "Risk and System Integrity Levels".

2.1   Integrity Levels

The four basic integrity levels defined in the IEC document range from the highest integrity (Class 4) to the lowest integrity (Class 1). At the lowest defined level (Class 1) simple fail safe techniques are all that is required, these imply a level of self test and monitoring and provided a separate control system is in place, a simplex safety system plc would be adequate.

At the highest level of integrity all measures need to be taken to eliminate dangerous failures, this includes multiple redundancy and diversity in configuration design as well as taking all appropriate life cycle measures.

The purpose of this paper is to provide guidance to enable Programmable Logic Controllers to be configured to provide integrity levels at the highest class of safety (Class 4) when applied to systems that have a known safe state.

3.   Safe State Systems

The types of configuration that will be described in this paper are primarily suitable for protection systems that have been inherently designed for fail to safety.
They encompass the majority of process plants and invariably have to meet the systems requirements of fail to safety, high availability and maintainability.

The combination of these requirements directs that the following features are provided without jeopardising safety:-

1)   Testability

2)   Repairability
3)   Configuration Modification (System Configuration Maintenance)
4)   Partial Plant Maintenance

and at the same time minimising the potential down times of the plant through maintenance of false trips.

3.1   Class 3 Safe State System

In all the configurations described it is assumed that the basic design criteria specified in IEC 65A WG10 Annex A are adopted to the recommended or highly recommended levels.

It can be seen by observing the above criteria that configuration shown in Figure 1 (Dual Redundant 2 out of 2 Voting) and Figure 2 (Triple Redundant 2 out of 3 Voting) will meet the requirements of Class 3. The advantages and disadvantages of each configuration and the higher level integrity limitations are described below.

3.1.1   Class 3 Dual Redundant 2 out of 2 Voting

The system configuration shown in Figure 1 would require that the inputs and outputs are regularly tested to ensure that no latent 'stuck on' conditions exist. Although one 'stuck on' fault can be tolerated, unless it is detected and rectified eventually two faults in the same channel would appear resulting in a potentially dangerous fault condition.

There would also be fundamental limitations defined with respect to repair and maintenance times, as it would need to be recognised that whilst one channel was non-operational, total reliance on the single fully functional channel would occur. This maintainability problem, which is of course also linked to testability (only one channel would be on-line and providing protection whilst the redundant channel was being tested) prevents this type of configuration being considered for the highest class of integrity.

The source software, although presumably with millions of hours of fault free operation, would almost certainly be identical in both systems. Similarly the application software, although designed using the finest techniques, would probably be identical in both systems. These potential common cause failure problems would also limit the level of integrity allowed for with this type of configuration, even assuming other redundant control paths were available.

3.1.2   Class 3 Triple Modular Redundant 2 out of 3 Voting

The configuration shown in Figure 2, if correctly designed, eliminates the majority of the testability/maintainability problems associated with the dual 2 out of 2 configuration, also availability is significantly increased as the potential for false trips due to overt faults is virtually eliminated. However, the common cause failure problem remains with respect to source and probably application software.

4. Moving Up the Integrity Class

The primary method of increasing the integrity level is to eliminate wherever practical the common mode failure mechanism. With respect to the system configuration, this invariably means adding diverse system to the safety system.

Certain assumptions at the system design level have to be made:

1) The HAZOP has been correctly completed

2) The basic safety design data is correct (cause and effects/fault schedules)

3) That no mistakes have been made in documenting the above

Although the clarification levels have been sharply defined as 1, 2, 3 and 4 in actual fact in each of the levels a range of configuration types would be acceptable, each individual configuration with a slightly better or worse integrity level than its peer, but nevertheless falling within the specified category.

The first type of integrity improvement I wish to consider remains still within the Class 3 classification, but puts the configuration firmly at the top end of Class 3.

4.1 Triple Modular Redundancy with Diverse Random Testing

In operational mode this configuration is identical to Figure 2, however, prior to the system being accepted the configuration shown in Figure 3 would have been operated for several hundred hours without discrepancies being reported.

The configuration shown in Figure 3 provides an additional diversely design test harness activator. In an example of this configuration known to the author, the unit termed SAPTU, Software Application Program Test Unit, is programmed directly using the prime cause and effects or fault schedules for the base data.

When operating, random pattern inputs are continually presented to the functional TMR unit and the output results are monitored for both compliance and discrepancies.

Diversity has been utilised in both the design of the test tool and the application programming providing a high level of confidence that common mode software design failures do not exist. It also provides a level of confidence that the base data is correct.

24

4.2　Class 4 Safety Systems

To provide a safety system classified to level 4 full functional and operational diversity does need to be provided. As in the other classes, a range of configuration will be acceptable, however, it is evident that at the lower end of the Class dual diversity is probably appropriate, where as at the upper end triplicated diversity or greater would probably be required.

With fail safe state systems it is improbable that diversity above the level of 3 would ever be required.

4.2.1　Class 3/Class 4 Mixed System

The system configuration shown in Figure 4 is a typical configuration utilised in the petrochemical industry to ensure both high availability and, where applicable, ultra high safety for fail safe state process plant. The addition of trip amps and logic provides High Integrity Protection for those parts of the plant that require Class 4 safety integrity coverage. Usually on a petrochemical plant the percentage of safety Inputs and Outputs needed to be protected to this level of integrity is small (less than 5% of all I/O), therefore the cost of lifecycle maintainability of providing this level of diversity is acceptable.

It should be noted that the triple modular redundant trip amps would normally be from the same manufacture and that each solid state leg would be manually tested (only one of the three legs would be testable at any one time).

Where High Integrity protection Systems (HIPS) circuits are applied it is extremely unlikely that start-up overrides or other such circuits would be needed or allowed on the small percentage of circuits covered by the HIPS diversity.

4.2.2　Class 4/Class 3 Mixed System

The system configuration shown in Figure 5 in many ways resembles that shown in Figure 4. However, the solid state trip amp circuits now provide an additional level of diversity being supplied by different manufacturers, hence increasing the integrity of the HIPS circuits.

4.2.3　Class 4 System Diverse TMR

Although logically the approach to provide diverse TMR would imply three diverse processing systems voting on inputs and outputs, the practicality of this approach for fail safe state systems in general precludes this configuration. A more practical configuration is that shown in Figure 6, which utilised dual diverse proprietary TMR systems. Using standard TMR systems currently available a diverse systems configuration can be provided that will not only maintain reasonably high availability due to the inherent fault tolerant design of each TMR systems, but also provide Class 4 integrity for all inputs and outputs.

As both TMR systems are programmed as Simplex programmable Logic Controllers, their lifecycle maintenance is manageable, also all faults are internally diagnosed and reported by each TMR system and all modules are repairable hot on-line.

The application programming for each system would of course need to be diversely implemented by separate teams based in separate organisations.

4.2.4   Class 4 Diverse TMR Plus

To increase the integrity of the system to even higher levels on certain Inputs and Outputs, the approach adopted in 4.2.1 and 4.2.2 can be utilised in conjunction with a diverse TMR system. This is shown in Figure 7, once again normally only a small number of Input and Output points would need to be configured through the solid state system. It is unlikely that using diverse trip amps would significantly add to the classification level of this configuration.

5.   Conclusion

There are many ways of achieving high levels of safety integrity through system configuration and in this paper I have attempted to indicate by example some of the more practical methods currently used, however, the configurations in this paper should not be considered as exhaustive of all possible fail safe state configurations.

The Safety Community, I am sure, awaits with anticipation the finalisation of the current draft documents which should provide strong guidance on the current and safe approaches to be taken for both personnel and capital equipment protection. For the safety industry to work to one initial international standard, will I am sure, improve overall safety awareness and increase the levels of actual safety integrity applied.

References

PES I and PES II   Health and Safety Executive

DIN VDE 801   German DIN Standard

IEC WG10 (Draft)   IEC

ISA SP84 (Draft)   ISA

"Risk and System Integrity Concept for   R Ball and D Reinert - Butterworth -
Safety Related Control Systems"   Heinemann

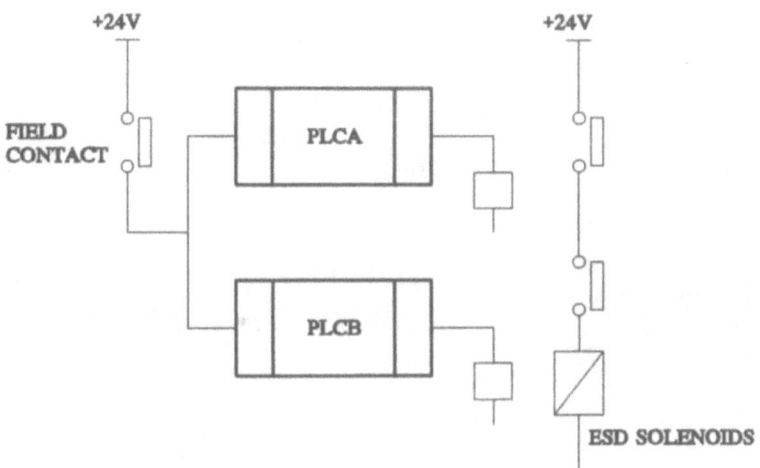

Figure 1

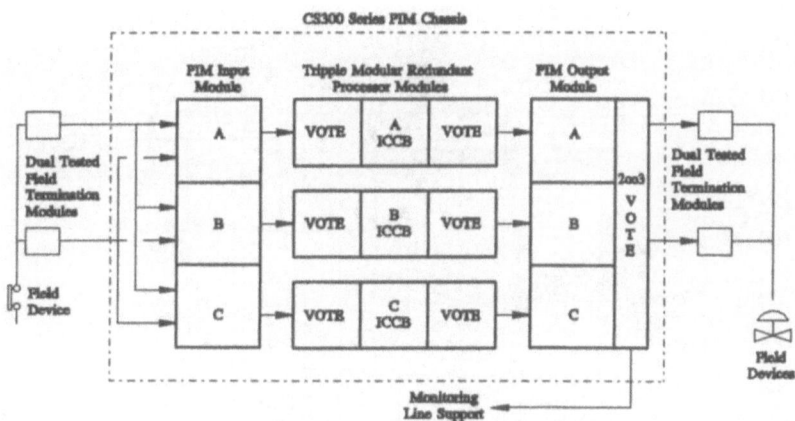

Figure 2

27

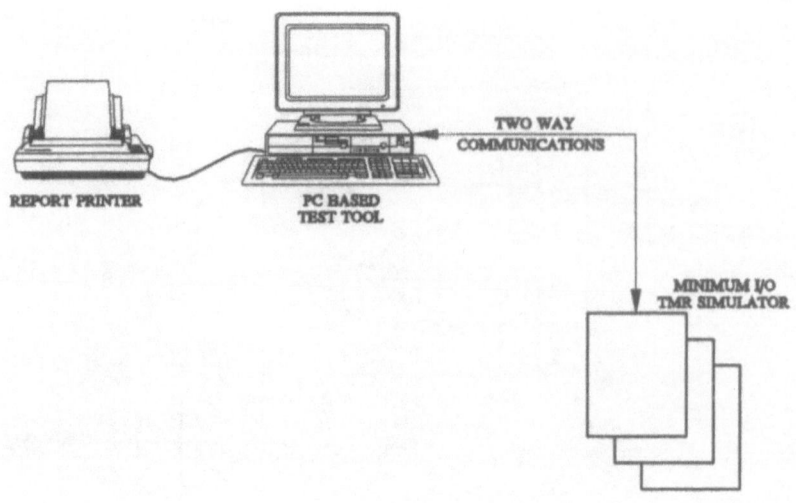

Figure 3

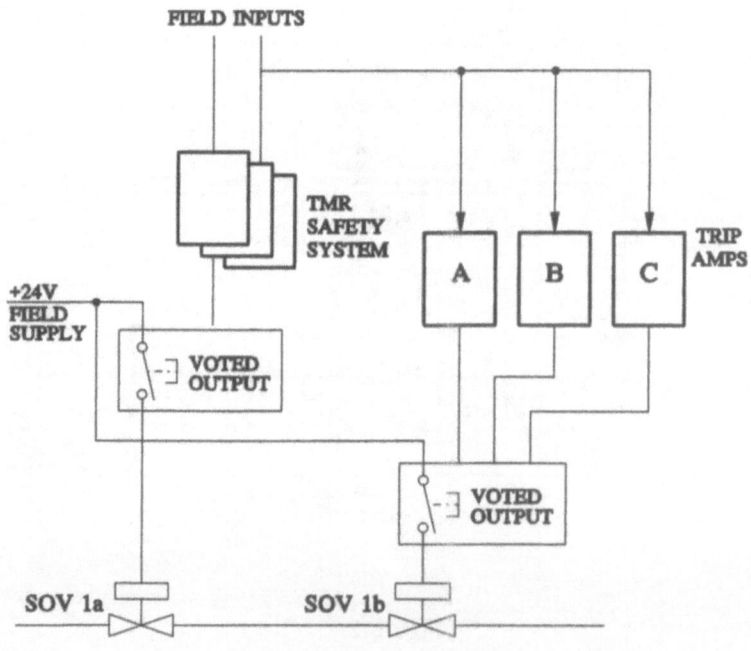

Figure 4

28

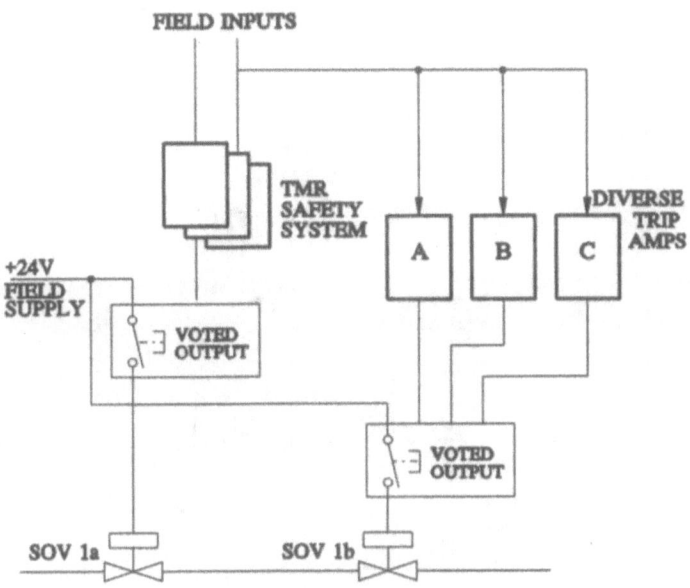

Figure 5

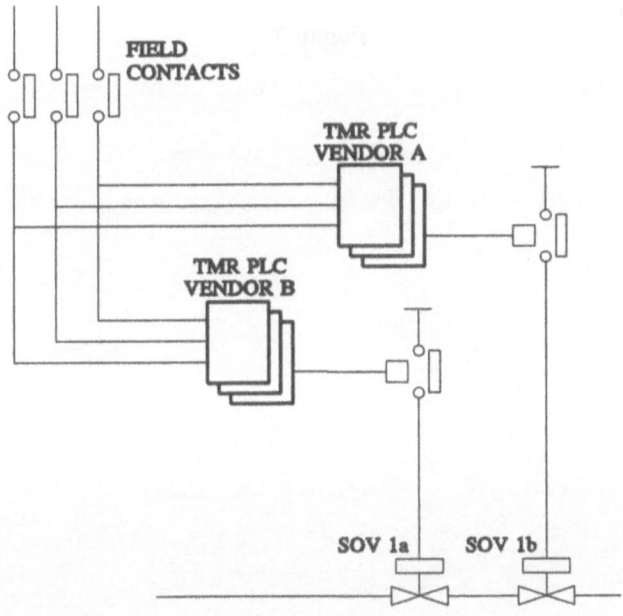

Figure 6

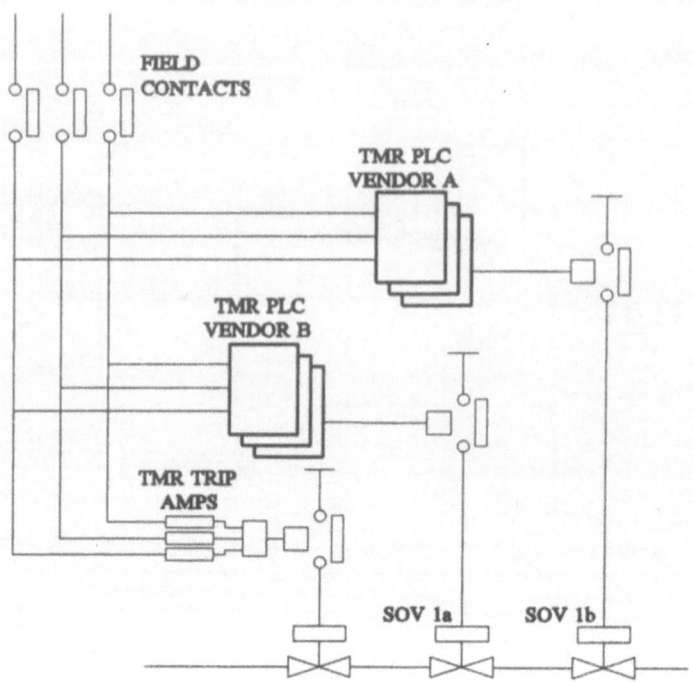

Figure 7

30

Maintainability
and its Influence on System Safety

Cris Whetton, CEC Research Fellow
University of Sheffield, Dept. of Mech. & Process Eng'g.,
PO Box 600, Mappin St., Sheffield, UK

1 INTRODUCTION

Maintenance is the totality of technical and administrative actions taken to retain an item in or restore it to its designed operating condition. Maintainability is the systematic assessment of the effectiveness of maintenance strategies and can have considerable influence on system safety; as such, it is more concerned with the design of the system than with the procedural or management aspects of maintenance. This paper introduces the concepts of maintainability as they relate to safety, concentrating on corrective maintenance and suggesting some of the ways in which these concepts can be applied. (In order to do justice to the subject, preventive maintainability is not covered.)

Much of what has been written about maintainability has been based directly upon the US Military Handbook 472 [1]; however, since this work refers only to the maintenance of electronic equipment under military conditions it must be used with great care since the maintenance model assumed may not be applicable to the system being studied. This paper presents a more general maintenance model and analyses the factors which contribute to a high degree of maintainability and hence to system integrity.

The maintenance of safety-critical systems can conveniently be studied with reference to the general characteristics of the system and those of its maintenance. Systems are generally of two types: those which support a **critical function** (E.g. life-support systems and railway signalling) and those with **inherent hazards** (E.g. chemical process plant and weapons systems.) These categories are convenient but not rigid; a system may belong to both categories or parts of it may belong to different categories. With regard to maintenance, it is also convenient to define two categories of system: those which are maintainable within their mission and those which are not. To be **maintainable-within-mission** (MWM), it must be possible to effect repairs while the system is performing its designed function (mission), even if this requires shutting the system down for a brief period (E.g. chemical process plant and railway signalling.) Otherwise, the system is **non-maintainable-within-mission** (non-MWM) and repairs are not possible until the mission is complete. This situation tends to apply to systems for which the term 'mission' has its everyday meaning, such as aircraft and space-craft but can also apply to some chemical reactions which cannot be interrupted without disastrous consequences. Use of these categories allows the maintainability requirements of the systems to be classified according to Table 1.

	Maintenance characteristics	
System type	MWM	non-MWM
Critical Function	High availability achieved by rapid maintenance, highly reliable equipment, redundancy, or a mix of all three.	High availability achieved by either high reliability or redundancy.
Inherent Hazards	Maintenance activities may be hazardous.	Protective systems may have to be classified as critical. Economics tends to determine whether reliability or redundancy is used to ensure availability.

Table 1 : Maintainability requirements by system type

All of the categories in Table 1 imply a high availability for the system, which can be achieved either by improving reliability or, often more economically, by improving maintainability. However, it is the combination of system type and maintenance characteristics that determines the possible strategies for achieving these goals.

1.1 THE MAINTAINABILITY CONCEPT

Maintainability is a measure of the effectiveness of maintenance strategies and is defined in BS 3811 [2] as:

> The ability of an *item*, under stated conditions of use, to be retained in or restored to a state in which it can perform its required functions, when *maintenance* is performed under stated conditions and using prescribed procedures and resources.

Unlike reliability, maintainability is rarely quoted as a probability but instead is usually quantified in terms of Mean Corrective Time[*] , $\overline{T}_c$ - the mean time required to restore the system to its original state after a failure has occurred and analogous to MTBF - Mean Time Between Failures. Such a number can be very useful for comparing alternative strategies though it must be realised that in most

---

[*] Some authors use the notation MTTR - Mean Time to Repair - but this can be confusing if no actual repair is involved. Others use MTTR to refer to Mean Time To Restore.

cases the number will have little absolute meaning. Mean Corrective Time, $\overline{T}_c$, is defined [1] by the equation:

$$\overline{T}_c = \frac{\sum_{i=1}^{n} T_{ci} \cdot \lambda_i}{\sum_{i=1}^{n} \lambda_i} \qquad ...\{1\}$$

Where: λ_i is the failure rate of the i'th repairable assembly
T_{ci} is the corrective time for the i'th repairable assembly
n is the number of repairable assemblies

To maximise availability, it is most economical [3] to minimise $\overline{T}_c$ but in all cases it is obviously desirable to minimise the resources spent on maintenance; to do this requires an understanding of the factors which go to make up T_C, the corrective time of a subsystem or assembly.

Another metric is the Mean Maximum Corrective Time, $\overline{T}_{c(max)}$. It is often assumed, especially for electronic systems, that repair times are log-normally distributed. In such a case, $\overline{T}_{c(max)}$ conventionally refers to the 90th or 95th percentile time; i.e. the time by which 90% or 95% of all maintenance tasks will have been accomplished. This time is of importance e.g. when planning 'suspension' strategies, such as described in paragraph 4.3.3, below.

Other metrics of importance are fault-coverage, the proportion of faults detected by the diagnostic system, and ambiguity, the ability to discriminate between faults having the same symptoms but different causes. These are discussed in detail below.

1.2 MAINTENANCE INDUCED FAILURES

One aspect of safety-critical systems that is often overlooked is the question of maintenance induced failures. Table 2, below, suggests that these are far more common than might be expected. While too much importance should not be attached to these figures since the authors cited are not necessarily consistent as to what they regard as maintenance errors, they suggest that there is a significant probability that a maintenance action will not be completely successful. Interestingly, none of the referenced papers consider equipment design as a contributory factor to maintenance error; nor does an HSE publication devoted to the subject [10], yet data given in that publication, when rearranged, suggests that of all maintenance incidents examined, 23% were attributable to the design of the equipment concerned. Hence, maintainability has an indirect effect on system safety in that any maintenance action must be done in such a way that it does not introduce new hazards into the system.

Description	%	Sample	System type	Ref.
Defective procedures	35	138	Unspecified process plant	4
Maintenance	40	921	Unspecified process plant	4
Defective maintenance	9	615	Japanese process plant	5
Poor maintenance	4	?	Elec. & pneumatic instruments	6
Maintenance errors	8	81	Refinery equipment	7
Faulty maintenance	30	?	Rotating machines	8
Human induced errors	66	?	NASA Space Shuttle	9

Table 2 : Incidence of maintenance-induced failures

This is clearly common-sense: if equipment is designed to be readily maintainable then there is greater likelihood that the maintenance task will be accomplished correctly. Rightly, design effort tends to be concentrated on the operational performance aspects of a system; unfortunately, rather than being included as a desirable aspect of system performance, maintainability often seems to sink towards the bottom of the designer's list of priorities.

1.3 Maintainability-induced Hazards.

Clearly, maintenance activities inherit the hazards of the system on which they are performed and such hazards are usually independent of whether the system is or is not safety-critical; however, certain aspects of maintainability can induce additional hazards. For example, if Built-In Test Equipment (BITE) is used, it must be remembered that a failure of the BITE can induce a system failure. Similarly, external Special Test Equipment (STE) and Maintenance Assist Modules (MAMs) must not only be designed so that their failure does nor bring down the system but also so that their use does not introduce 'sneak' conditions [11], leading to transient operational failures. It is thus vitally important that the design of such ancillary systems be co-ordinated with that of the main system and that they are also included within the scope of the system safety analysis. Finally, it must be remembered that the inclusion of BITE will adversely affect the system's reliability because the component-count will be increased; thus, BITE should always be examined in an availability analysis, to ensure that the improvement in maintainability offsets the loss of reliability.

2 A FAILURE MODEL

To understand how maintainability influences the integrity of a safety-critical system, it is first necessary to examine possible responses to equipment failure. Using the IEC terminology that Failure is a deviation from specified service and a Fault is the cause of a failure, the sequence of events arising from a fault (regardless of system type or maintainability characteristics) can be as shown in Figure 1, where three choices are offered in the event of a failure: Do nothing; Compensate; or Correct. (Note that for <u>any</u> action to be possible, the failure must be detected and the fault located before the desired action can be selected.)

2.1 Do nothing.

The failure may not be safety-related, having negligible consequences and it may be most economic to defer any maintenance until a scheduled shutdown occurs or until sufficient failures have occurred to make a maintenance action worthwhile. Such a situation is typical of light-bulbs failing in a large room: failed bulbs are not replaced as they occur but when, say, 20% of bulbs have failed.

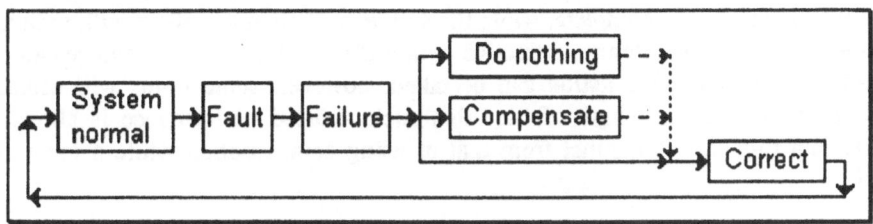

Figure 1 : Maintenance options

Such an action can also apply to the failure of safety-critical items if the item is backed up by a redundant element.

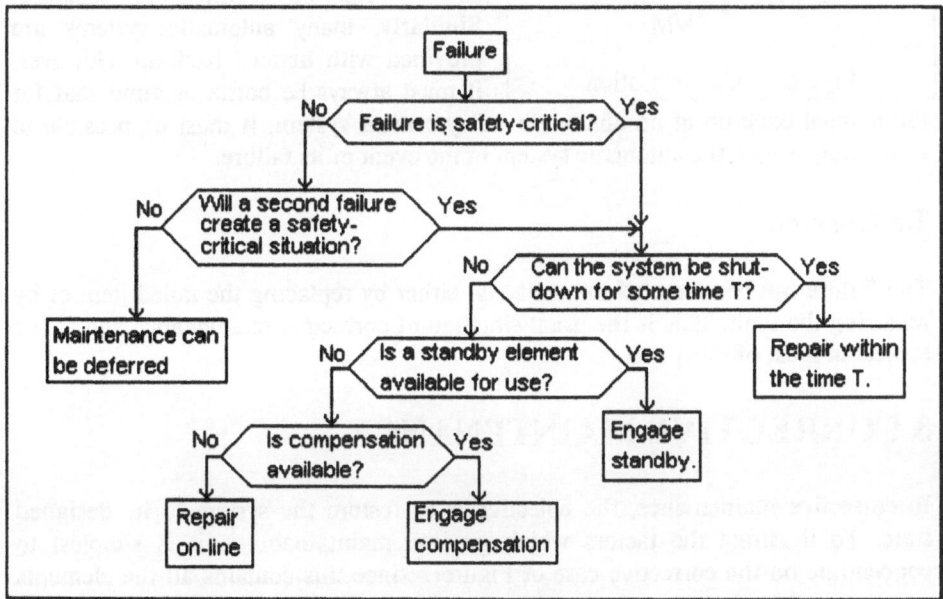

Figure 2 : Choice of strategies.

However, such a choice must be made with care, as Figure 2 shows. While the figure is only meant to be illustrative of the ways in which choices can be made amongst these three options, it suggests that maintenance should only be deferred if the failed item is not itself safety critical <u>and</u> if a second, subsequent failure will not create a hazard. 'Do Nothing' is often the only option for a non-MWM system and it may be necessary to resort to active parallel redundancy to ensure safety. A common example of this being the four-engined airliner which can fly and land with only two working engines.

2.2 Compensate.

The failure may be such that existing equipment can be used to compensate for or 'work around' the failure, delaying maintenance until a scheduled shutdown occurs or until the mission is complete, while the system still operates, albeit with reduced performance. This is often an attractive option if some loss of performance can be tolerated until corrective action can be taken; however, some limits will usually need to be set on the length of time that such reduced performance is allowed. (Note that this case is distinct from that of using compensation while a repair is made.)

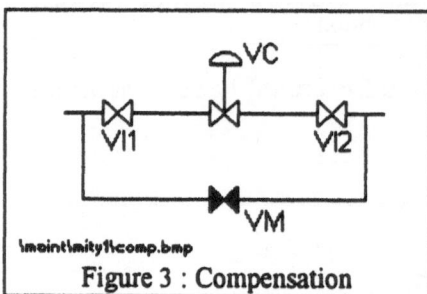

Figure 3 : Compensation

Figure 3 shows a common example of compensation. If the control valve VC fails, it can be isolated by the valves VI1 and VI2 and control can be maintained manually by the valve VM until such time as VC can be repaired or replaced. Similarly, many automatic systems are provided with manual back-up. However, it must always be borne in mind that for the manual back-up of an automatic safety-critical system, it must be possible to completely isolate the automatic system in the event of its failure.

2.3 Correct.

The failure can be corrected immediately, either by replacing the failed item or by repairing the fault. This is the usual situation of corrective maintenance and is the subject of most of this paper.

3 CORRECTIVE MAINTENANCE

In corrective maintenance, the objective is to restore the system to its designed state. To illustrate the factors which go into maintainability, it is simplest to concentrate on the corrective case of Figure 1 since this contains all the elements common to the others.

The corrective case is shown in slightly more detail in Figure 4.

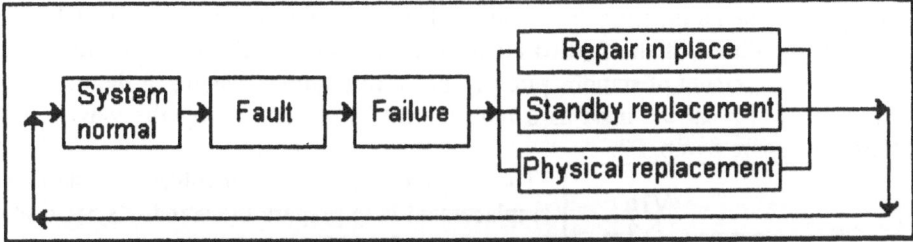

Figure 4 : Corrective actions

In general, there are three ways of restoring a system to its original state by corrective actions: **Repair in place**; **Standby replacement**; and **Physical replacement**. Each applies in different circumstances and has different characteristics.

3.1 Repair in place.

This applies to assemblies which cannot economically be removed or where a faulted component can safely be replaced under field conditions. In the latter case, it can involve the use of compensation, as illustrated in section 2.2, to allow the system to operate while repair takes place.

First, the system must be shut down or otherwise rendered safe to repair. Next, the assembly to be repaired must be prepared (drained, purged, etc.) and the necessary parts, personnel, and tools must be assembled; these two activities can take place in parallel. Then, the faulted component is either repaired in place (e.g. welding up a crack) or removed and then replaced by a good component. Once the repair has been effected, it must be checked. Finally, the system is re-started. The tasks involved are shown diagramatically in Figure 5.

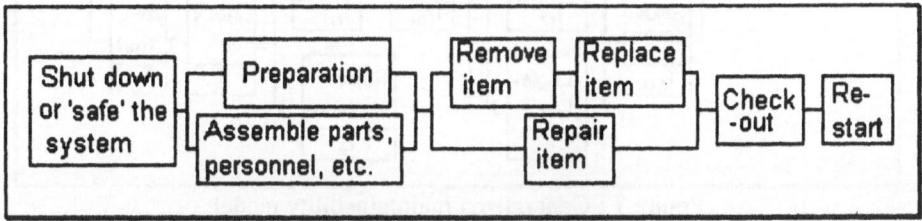

Figure 5 : Repair in place.

This can apply to the case of, say, a pump seal replacement when there is no standby pump available. The faulted part - the pump - remains in place and the process is shut down until the repair has been made.

3.2 Standby replacement.

This applies when a functionally identical element is held permanently available to take over the duties of the faulted element. Such a case is illustrated by the very common arrangement of standby pumps, shown in Figure 6, where, in the event of failure of P1, P2 can be used as a replacement by re-configuring the valves and starting the spare pump, P2.

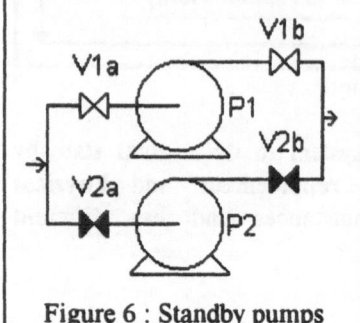

Figure 6 : Standby pumps

The sequence of events involved in standby replacement is to prepare the standby system for operation (if necessary); change over to the standby system; and verify that it is operating properly.

3.3 Physical replacement.

In physical replacement, the failed system, rather than the failed component, is replaced with one which is functionally identical; repair of the fault takes place elsewhere - usually in the local workshop, a central depot, or at the manufacturer's facility. The sequence of operations is identical to those of *Repair in place*, shown in Figure 4 but the details are different.

4 MAINTENANCE ACTIVITIES

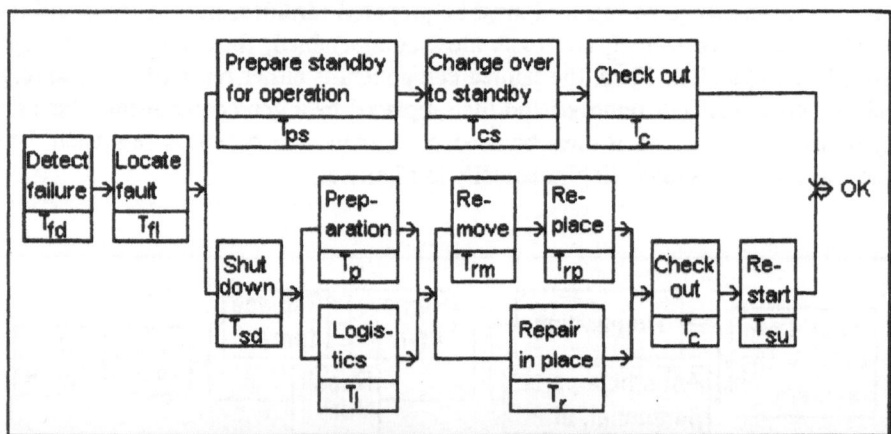

Figure 7 : Generalized maintainability model

Each of the activities in each of the strategies outlined above consumes different resources - especially Time - and has different implications for safety; consequently, the study of these activities and their times is essential to the qualitative and quantitative evaluation of maintainability. The activities shown in Figures 4 and 5 may be combined into one diagram such as Figure 7, from which twelve fundamental operations and their associated times can be identified; they are listed in Table 3, below. The sum of a particular combination of these times gives the Corrective Time, T_C, in the equation for Mean Corrective Time, $\bar{T}_c$, {1}.

Time	Activity
T_{fd}	Detect a failure
T_{fl}	Locate the fault which caused the failure
T_{ps}	Prepare a standby system (if any) for operation
T_{cs}	Change over from the failed system to the standby system
T_{sd}	Shut down system or otherwise render safe while correction takes place.
T_p	Prepare the failed system for repair
T_l	Assemble the necessary materials, tools, personnel, etc.
T_{rm}	Remove the failed item.
T_{rp}	Replace the failed item.
T_r	Repair the failed item.
T_{co}	Check out the repair or replacement.
T_{su}	Re-start the system.

Table 3 : Maintenance activity times

For example:
(a) A standby system, such as Figure 6.

$$T_c = T_{fd} + T_{fl} + T_{ps} + T_{cs} + T_{co} \qquad ...\{2a\}$$

(b) An in-place repair, such as welding a crack in a pressure vessel.

$$T_c = T_{fd} + T_{fl} + T_{sd} + \left(T_p \underset{max}{|} T_l\right) + T_r + T_{co} + T_{su} \qquad ...\{2b\}$$

(c) Physical replacement of a complete item, such as a circuit module or power supply.

$$T_c = T_{fd} + T_{fl} + T_{sd} + \left(T_p \underset{max}{|} T_l\right) + T_{rm} + T_{rp} + T_{co} + T_{su} \qquad ...\{2c\}$$

An examination of the factors which determine the individual activity times can be very revealing, both from the point of view of minimising the time and from that of accomplishing the activity safely and reliably; in the following section these are examined in detail. Note that the times to detect a failure and locate a fault are common to all possible sequences of actions; because these factors are so important, they are treated last.

4.1 Standby Preparation : T_{ps}

This is the time required to prepare a standby system for operation. In maintainability analysis, as with reliability, it is usual to recognise two types of standby system: *hot* and *cold*. In the case of hot standby, the spare item is operating but not carrying any load and the preparation time is essentially zero.

In the case of cold standby, the spare item is not operating and must be brought on line in a co-ordinated fashion with the changeover activity. Two conditions must also be satisfied :

1 The item must be known to be working before changeover.
2 The item must be in the appropriate state before changeover.

The second condition is complicated by the fact that the failed element may induce an invalid state in the rest of the system. Consequently, care must be taken to ensure that the failed item is completely isolated from the system, so that the latter's state may be properly determined before changeover occurs.

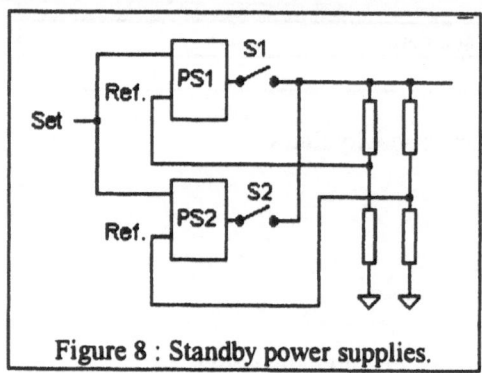

Figure 8 : Standby power supplies.

Consider Figure 8 where power supply PS2 forms a redundant standby to PS1. If PS1 fails, it must be disconnected from the load so that it does not affect the output of PS2 as it comes up to voltage; this requires the voltage reference to be down-stream of the switching. However, such an arrangement may introduce an additional hazard because PS1 and PS2 are permanently connected to their reference points, possibly exposing maintenance personnel to unexpected voltages unless additional precautions are taken. Such conflicts between operational and maintainability requirements are not unusual, especially in the practical implementation of standby systems.

4.2 Standby changeover : T_{cs}

This is the time taken to change over from the failed system to a standby system which is ready to assume a load. There will always be a finite T_{cs}, even in the case of hot standby ($T_{ps}=0$) with automatic changeover. For safety-critical systems, the timing and sequence of the changeover should be examined in detail, especially since apparently simple tasks may hide subtle complications, such as in the example of Figure 8. Techniques for analysing maintenance tasks are discussed elsewhere [3, 12, 13], and are not repeated here.

4.3 Safe or Shutdown : T_{sd}

This is the time required to render the overall system safe before actual repairs or dismantling can begin. Regardless of system type, there are two possibilities : either the system is shut down completely while repair takes place; or the system is reconfigured so that repair may take place safely while the rest of the system continues to operate, possibly with reduced performance.

If shutdown is the chosen option it must be designed so as to be accomplished safely. While this may be a minor problem for some systems (e.g. those which are primarily electronic or electromechanical) shutdown of, say, a chemical process plant may take days and require a complex and carefully controlled sequence of operations. Furthermore, a shutdown will be followed by a re-start and experience suggests that for many systems, start-up is the most dangerous part of operation.

Consequently, wherever possible, shutdown should be avoided on large and complex systems; however, this requires that the system is designed to allow safe maintenance while still operating. Assuming that all the possibilities of redundancy have been exhausted (usually for economic reasons) the three options remaining as alternatives to shutdown are : Compensation; Bypass; and Suspension.

4.3.1 Compensation
Compensation, as already discussed, involves replacing the faulted section with some alternative, usually of lower performance, while repair takes place.

4.3.2 Bypass
Bypass involves losing the facility of the faulted section for the duration of the repair. I.e. the system operates at lower performance and no attempt is made to compensate for the loss. As an example, the faulted de-humidifier of a submarine's air conditioning system would simply be bypassed while a repair was made, knowing that the crew could tolerate the resulting discomfort for some time.

4.3.3 Suspension
Suspension is generally most applicable to 'production' systems and can best be

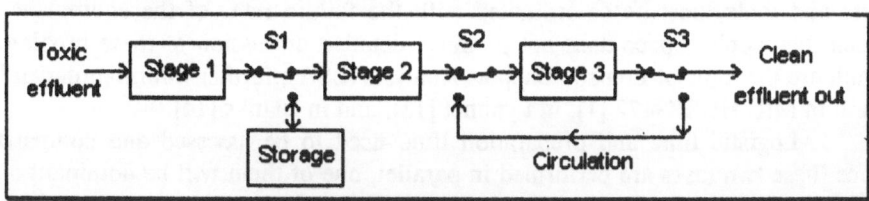

Figure 9 : Suspension

illustrated by an example. Figure 9 shows, say, a water-treatment plant for processing toxic effluent. If Stage 2 of the system fails, material from Stage 1 can

41

be diverted to storage while repairs are made. However, it may be necessary to keep material circulating through Stage 3, so as to maintain process temperatures, etc. Such schemes need to be designed with great care : maximum flow rates and the time taken to repair Stage 2 under worse-case conditions (i.e. $\overline{T}_{c(max)}$, as discussed in paragraph 1.1) would need to be known accurately before the size of the required storage could be determined. In general, the suspension option is less to be preferred than either compensation or bypass.

4.4 Preparation for repair : T_p

This is the time to prepare the failed item for repair and is distinct from T_{sd} which refers to the overall system. It covers such activities as draining process fluids from lines and vessels and steaming-out to remove flammable or toxic vapours, discharging high voltages, and allowing high-speed rotating machinery to come to a stop. This task can usually be accomplished in parallel with the logistic time, T_l. For maintenance to take place safely, the equipment must be designed with such preparation in mind; for example:

- It must be possible to isolate the item from the rest of the system and positively indicate that the isolation is effective.

- If energy, in any form, can be trapped within the isolated item, then safe means must be provided for its release and a positive indication provided that such energy is no longer present.

4.5 Assembly of materials, tools, etc. (Logistic time) : T_l

In any maintenance activity, it is unreasonable to assume that the necessary materials, equipment, and personnel are instantly available. Unless spare parts are held at the point of likely failure (as is commonly done on some military systems) they will have to come from some other location. This topic is discussed in greater detail in another paper by the author [3].

This activity is not necessarily a trivial task and the availability of spare parts and tools must be co-ordinated with the failure rates of the equipment to which they apply. Space does not permit a detailed discussion of these problems, which are the topic of Logistic Support Analysis (LSA); further information can be found in MIL-HDBK-472 [1], in Lyonnet [13], and in Blanks [14].

Logistic time and preparation time need to be assessed and compared. Since these two tasks are performed in parallel, one of them will be dominant and efforts at reducing the time (without compromising safety) can then be concentrated on the dominant task. Since preparation time is often determined by safety considerations, logistic time, if dominant, offers the most scope for reduction. For example, the time can be reduced by designing equipment to avoid the need for special tools.

It is sometimes argued that logistic time should not be counted as a factor of maintainability since logistic performance is not determined by equipment design and because maintenance and logistic support are usually performed by different organisations. The following examples show why logistic time should be included:

Tools. The assembly of the necessary tools is generally considered a logistic task. Tools (which includes anything from a screw-driver to a tower crane) generally come from three sources:

(i) The maintainer's personal stock. These are typically common hand tools and simple test equipment such as pressure gauges and multimeters. The logistic time for such items is negligible (provided that they have not been lost or forgotten).

(ii) From a local store. This includes bulky tools or ones which are not regularly used by each individual maintainer but which are used sufficiently often that their retention on-site is economically justified. Typical examples might be a logic analyser, a pipe-bending machine, or a set of imperial spanners on a site where the majority of fasteners are metric. The logistic time for such items is determined by geography and the procedures for booking the item out of stores.

(iii) External to the organisation. Some items are used so rarely or are so specialised that they are not held by the organisation but must be obtained (e.g. hired) from external sources. A typical example is heavy lifting equipment, for which the logistic time may be measured in days, or longer.

Materials. Collecting together the materials, spare parts, etc. needed to perform a repair or replacement is a logistic task in the classic sense. Like tools, materials generally fall into three classes of accessibility:

(i) The maintainer's personal stock (which may be replenished daily from the central stores of (ii), below). This typically includes items such as fuses, small fasteners, wire, and small gaskets. Logistic time for such items is negligible.

(ii) From a local store. This includes bulky items, ones which are not regularly used, or ones which are of high value. Typical examples might range from a complete pump to a microprocessor. The logistic time for such items is determined, as for tools, by geography and the procedures for booking the item out of stores.

(iii) External to the organisation. Some items are used so rarely that they must be bought from external sources, as and when they are required. A typical example is a chemical reactor vessel, for which the logistic time may be measured in months.

In each of the above cases, design of the equipment determines whether or not the item is needed. If it is not needed, then it incurs no logistic time and it is in this sense that design can influence logistics. If all the fasteners used on the system are to the same standard then the need to draw special tools from stores is avoided. At the other extreme, if the design is such as to allow for the maintenance of heavy equipment *in situ*, the need for lifting equipment from outside sources can be avoided.

To give a concrete example of such factors, the author was once involved in an audit of the maintenance activities of a large US teaching hospital. The hospital had a large maintenance department, with about twelve engineers and technicians and possessed about thirty specialised beds for treating burns victims and spinal injuries. These beds had originally been electrically operated but had recently been replaced with newer ones which were mainly hydraulic. Since the hospital had neither the equipment nor the personnel to maintain these hydraulics, they were dependent upon the supplier who was located about two thousand miles away. Consequently, about four of these beds were typically inoperable at any time, sometimes for periods as long as a month, compared to the older system where they were rarely unavailable for more than twenty-four hours.

4.6 Removal : T_{rm}

This is the time required either to remove the entire failed unit or to gain access to and remove the failed component. It can be estimated on the basis of experience, from maintenance records, or by analysing the elementary activities which form the removal task. There are several ways to do this: one, rather complicated method is given in MIL-HDBK-472 [1]; a less complicated method is given by Lyonnet [13]; and a simple but effective scheme is offered by the author [3, 12].

4.7 Replacement : T_{rp}

This is the time required either to replace the entire failed unit or to replace the failed component and any parts which had to be removed to gain access to it. It can be estimated and analysed in the same way as T_{rm}, above, using the same techniques.

It is not safe to assume that replacement is the exact opposite of removal, either in time or in the tasks involved. Bolts on equipment are often corroded or painted over and may be very difficult to remove; it is thus reasonable to assume that their removal time will exceed that for replacement. However, bolts must sometimes be replaced and torqued in a particular sequence; in which case, replacement time may exceed removal time!

In estimating T_{rm} and T_{rp} by the methods referenced above, use is often made of standard times for elementary actions; for example, MIL-HDBK-472 gives tables of such times. It must be stressed that <u>the standard times of MIL-HDBK-472 are only applicable to electronic and electromechanical systems under conditions of military</u>

maintenance. In synthesising times from elemental actions, experience and operational data are the most reliable guides. It is also important to remember that, when comparing design options, consistency is more important than the absolute values of the times assumed.

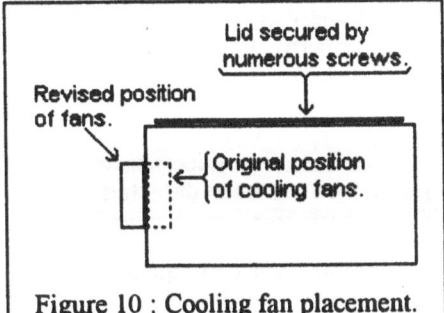

Figure 10 : Cooling fan placement.

Obviously, both T_{rm} and T_{rp} vary greatly with the design of the equipment and simple design changes can have significant effects on the task times. Figure 10 shows an example from the author's experience. A computer was cooled by three fans which were mounted inside the case. Access to the fans was via the case lid which was secured by numerous screws and a woven gasket so as to meet TEMPEST criteria. The fans had the highest failure rate of any replaceable component in the system but their remove/replace times were dominated by the time taken to remove the case cover. Transferring the fans to the outside of the case (and shortening the case so as to remain within overall volume) reduced the mean corrective time by about 30%, and allowed easier visual inspection of the fans.

4.8 Repair : T_r

This applies to cases where either the repair is made *in situ* (e.g. welding a cracked pressure vessel) or when the item is removed, taken away to a workshop, repaired, and returned to its original location. In the latter case, transport times must be included. In both cases, the repair must be analysed from the point of view of the actual tasks involved. In every case, it should be questioned whether or not repair in place is the best policy. Often, there is no option: it is difficult to move a 100 tonne vessel back to the workshop. However, with smaller items, and considering that repairs made in the workshop tend to be of higher quality than those made in the field, it should not be assumed that repair in place is always the best method.

4.9 Check out : T_c

All tasks must be checked before the equipment is returned to duty. As noted in the introduction, maintenance induced failures are all too common - at least, on chemical process plant - suggesting that this task is not always carried out as conscientiously as it should be.

Ideally, the check should be independent but this is not always possible. One possible compromise is for the maintenance procedure to contain a checklist of items to be verified, which is first completed by the person(s) responsible for the work and then briefly verified by a third party. A checking procedure should always form part of a permit to work system. In electronic and electromechanical systems, BITE can be co-ordinated with the checkout task.

4.10 Re-start : T_{su}

Like shut-down, this is usually determined by safety and operating requirements and little can be done to reduce it. The best strategy, as discussed in connection with shutdown, is to minimise the problem by designing to avoid a complete shutdown, leaving the system in a safe state while corrective action takes place.

4.11 Failure detection : T_{fd}

No corrective process can begin until a failure has been detected.

This simple statement is easily forgotten but has important consequences. For a failure to be detected, at least three things must happen:

The failure must produce a detectable effect.

The effect must be detected.

The operator (or an automatic system) must recognise the detection.

Consider the (extreme) example of Figure 11, which shows a storage tank with two level alarms, one high - LAH and one low - LAL:

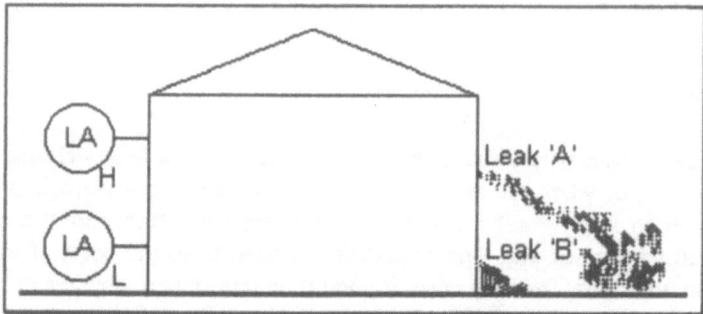

Figure 11 : A leaking tank

If a leak occurs in the storage tank at point 'A' then the contents of the tank will leak out until the level falls to that of the leak and, lacking any other instrumentation or visual observation of the problem, the leak will not be detected. Similarly, if the level in the tank always remains below that of the leak, it will not be detected. Clearly, the instrumentation is inadequate to detect this particular failure.

If the leak occurs at point 'B', there will eventually come a time when the low level alarm will trip and the existence of a leak can be inferred. In this case, it can at least be said that the average time taken for the failure (the leak) to be detected is the average time for the fluid to fall from a maximum at LAH to a minimum at LAL - provided that the size of the leak is known!

This simple example clearly shows that detecting failures is not as straightforward as it may seem, especially if an estimate of the detection time is required. In fact, failure detection cannot begin to be analysed until the equipment's

46

potential fault modes are known. This is discussed at greater length in section 5 on Fault Coverage.

4.12 Fault location : T_{fl}

This is the time required to locate a fault, given that a failure has been detected. Even though a failure may have been successfully detected, neither the corrective process nor compensation can proceed any further until the location of the fault has been identified. Figure 12 illustrates this problem.

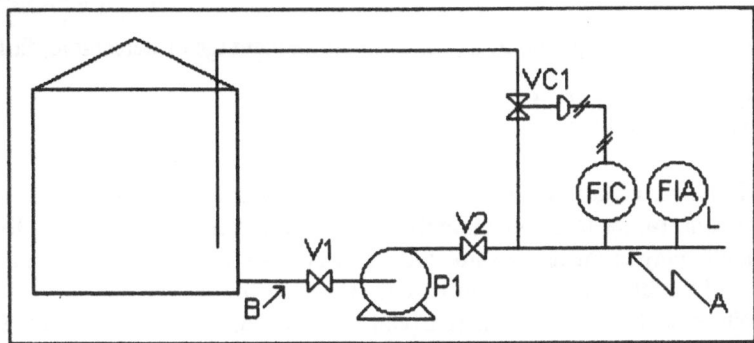

Figure 12 : Fault location

The failure in Figure 11 is low flow which is correctly, and almost immediately, detected by the flow alarm, FIAL. However, the (partial) set of possible faults which could produce this failure is:

Guillotine pipe fault at 'A'	Guillotine pipe fault at 'B'
V1 failed closed	V2 failed closed
VC1 failed open	VC1 forced open by FIC failure
P1 not operating	Tank empty

In this case, the instrumentation alone is not sufficient uniquely to locate the fault: the fault location is **ambiguous**.

The concept of ambiguity is important to Failure Detection and Fault Location (usually referred to collectively as Fault Coverage). If there is a one-to-one correspondence between failures and faults then there is no ambiguity. Ambiguity is discussed in more detail later in this paper and elsewhere [3].

There are other difficulties with fault location: faults may be referenced either to the failure that they produce or to the item of equipment that they occur on. If the main concern is safety then hazards are related to failures and interest centres upon what faults give rise to a given failure and how quickly they may be located and corrected before the hazard develops. On the other hand, if interest is primarily in effective maintenance then the concern is with how well the faults of a particular item of equipment may be located; i.e. the faults are referenced to the equipment, rather than to the failure. Since this paper is principally concerned with

47

the effects of maintainability on safety, this latter metric is not developed here but is discussed elsewhere [3, 12].

5 FAULT COVERAGE

Fault coverage is concerned with the two different but related concepts: the detection of failures and the location of faults; factors which are of great importance to maintainability and its effect on safety. The analysis centres mainly around the instrumentation* of the system.

Failure coverage is measured as a proportion or percentage of detectable failures:

$$d = \frac{\text{Number of detectable failures}}{\text{Total number of failures}} \quad ...\{3\}$$

The total number of failures is best obtained from the system (or subsystem) FMEA (though it can, rather less conveniently, be obtained from a HAZOP) recalling that several faults may produce the same failure. For example, if seven failure types are identified of which only three can be detected then the failure coverage is 3/7 or 43%. While such a calculation is trivial it does indicate, in this case, that the system is unsatisfactory though unless two or more different schemes are being compared, the actual value of the detection metric is of little interest. What the analysis has required that is important is to examine the instrumentation aspects of the FMEA in a systematic manner, as described elsewhere by the author [12] and by Blanks [14]. The aim should be to achieve 100% fault detection for those faults with the highest probabilities of occurrence and the highest potential hazard severity.

As noted above, fault location is concerned with finding the fault that caused the failure; it is complicated by the fact that several different faults can lead to the same failure. This leads to the concept of 'ambiguity', posing the question: Can the instrumentation uniquely identify all of the faults which can lead to a given failure? A quasi-formal definition of ambiguity is:

> For a given failure f, if the number of faults which can give rise to
> that failure is N_f then the ambiguity $\aleph^{**}$ is given by $\aleph = N_f - 1$.

The first step in assessing the level of fault location is to identify the ambiguity of

---

* Instrumentation is here used for any means of monitoring system condition and performance and does not necessarily refer to physical instruments such as pressure gauges, volt-meters, etc.
** There is no accepted symbol for ambiguity. 'A' is already used for Availability and α is used for modal fraction in FMEA, and $\aleph$ (aleph) happens to be on the author's word-processor.

48

each failure and to list them first in descending order of ambiguity, secondly, within each ambiguity band, in descending order of rate of occurrence. For example, as in Table 4.

Failure ID No.	Description	Ambiguity	λ
11	Low oxygen content	5	13.2
27	High temp., primary air	5	11.6
34	Low temp., primary air	5	10.3
9	Loss of refrigerant	5	3.2
14	High refrigerant temp.	4	26.5
91	Regenerator overpressure	4	0.6
90	Regenerator overtemperature	3	21.2
4	Low air flow	3	10.1
5	Low air pressure	3	9.6
etc.	etc.	etc.	etc.

Table 4 : Failure ranking by ambiguity

Table 4 suggests the order in which the failure location problems should be tackled; obviously, the higher the ambiguity and the higher the rate of occurrence, the more effort should be spent on providing instrumentation to resolve the ambiguity. This ranking method can be modified by including the hazard category. A separate ambiguity ranking is produced for each hazard category and the highest severity faults are resolved first. If hazard categories are in ascending order of severity, i.e., Category I is the most severe, a good rule of thumb to follow is: **The ambiguity should never exceed the hazard category**.

As noted above, a major restriction on maintainability is the time T_{fl} taken to locate a fault; this is dependent upon the level of instrumentation, the skill of the personnel, and the failure ambiguity. A structured method of estimating fault location time is suggested by Lyonnet [13], under the title 'arbre d'entretien' (Maintenance Tree). An improved version is given here. There are two different measures of fault location time:

T_{flf} - the time taken to locate the faults responsible for a particular failure.

T_{flu} - the mean time taken to locate all the faults associated with a particular unit or other hardware.

Of the two, T_{flu} is the more important for maintainability prediction, as described elsewhere by the author [3, 12] while T_{flf} is of interest for safety and in planning maintenance procedures and so is considered here in some detail.

5.1 Estimating T_{flf}.

The procedure is shown in Table 5.

Failure: Regenerator overpressure					
Ambiguity = 4					
Fault ID No.	Description	λ	T_l	ΣT_l	$\lambda \Sigma T_l$
031	Filter blockage	25.2	5	5	126
135	Controller failure	19.4	5	10	194
137	Pump overspeed	8.2	7	17	139.4
006	Column flooding	5.5	10	27	148.5
194	Steriliser blockage	0.3	15	42	12.6
		58.6			620.5

Table 5: Estimating T_{flf}

$$T_{flf} = \frac{\sum_{i=1}^{N}\left(\lambda_i \sum_{j=1}^{i} T_{i_j}\right)}{\sum_{i=1}^{N} \lambda} = \frac{620.5}{58.6} = 10.6 \quad \text{minutes}$$

In Table 5, the faults contributing to the failure 'Regenerator overpressure' are listed in order of their rate of occurrence. Next, the time to locate each fault T_l is listed. The fifth column is the cumulated time. It is assumed that the fault with the highest probability is tested for first, then the next highest, and so on. Thus, the time accumulated is that for the preceding faults, plus the time to locate the fault which is found. The sixth column is the product of the cumulated time and the rate of occurrence of the fault. The mean time is obtained by dividing the sum of the weighted times by the sum of the fault rates, or the occurrence rate for the failure.

5.2 Order of testing.

In the previous discussion, in both Tables 6 and 7 the faults were ranked in descending order of rate. Because of the accumulation of test times, this sequence does not necessarily give a minimum for the mean time. There is no analytical method for determining the minimum but computer programs based upon the 'exchange sort' algorithm are quite effective. As a rule of thumb, if the number of tests is less than ten, then testing according to descending failure rates will be with about 10% optimum. If the number of tests exceeds ten, then an optimisation program should be used to determine the best order.

6 DESIGN FOR MAINTAINABILITY

Checklists are often used for rating and ensuring maintainability at the design stage [1, 14]. However, while recognising that they are a useful summary of accumulated experience, the author places no great faith in design checklists because few designers have the patience and humility to apply them correctly. Rather than repeat such lists here, Tables 7-10 correlate design features which affect maintainability with the maintainability factors discussed in section 4, above. To keep the tables of manageable size, the maintainability factors are keyed as in Table 6.

a	Detect	e	Prep. for repair	i	Repair
b	Locate	f	Logistics	j	Checkout
c	Prep. for changeover	g	Remove		
d	Changeover	h	Replace		

Table 6 : Key to Maintainability Factors, Tables 7-10

Within the tables, factors which are closely linked are marked by •, those that are loosely linked are marked by ×, and those with little or no influence on each other are left unchecked. Note that start-up and shut-down are not considered in these tables since they tend to be dominated by operational factors.

The first table, Table 7, shows the influence of the various factors which may be grouped under the heading of Functional Testability. The importance of fault detection and failure location was discussed in section 5; however, as Table 7 shows, these factors extend beyond the detection and location of faults. The need for special test equipment influences logistic time. The presence of test points and BITE also affects preparation for changeover or repair because of the need to ensure a safe system state before proceeding with these activities.

Design factor	Maintainability factor									
	a	b	c	d	e	f	g	h	i	j
Functional testability	•	•	×		×	×				•
Fault location	×	•				×				×
Visual display of malfunction	•	•	×		·	×				•
Fault & operation indicators (BITE)	•	•	×			×				•
Identification of test points	×	•								•
Ease of testing in place	•	•	×			×				•
Use of standard test equip't	×	•	×		×	•				•
Special test equipment & MAMs	•	•	×			•				•

Table 7 : Functional Testability.

Similarly, BITE will have a strong influence upon checkout activities, especially where it can be used to relieve the maintainer of complex tasks and sequences.

Design factor	Maintainability factor									
	a	b	c	d	e	f	g	h	i	j
Physical features	×	•	×	•	•	•	•	•	•	•
Accessibility		•	×	•	•		•	•	•	•
Ease of removal & replacement			×	×	×	•		•	•	•	×
Ease of handling			×	•	×	•	•	•	•	•
Use of standard parts						•	×	•	•	×
Use of standard tools & test equip't	×	•	×		×	•	•	•	•	•
Ease of adjustment			×	•	×		×	•	•	•

Table 8 : Physical Features

Table 8 shows the influence of what can loosely be described as 'physical' features and amongst these, the two most important general areas are accessibility and standardisation. Accessibility is always compromised by performance requirements; e.g. weight or space restrictions may call for tight packing. As a general rule, where equipment must be packed tightly, the higher failure rate items should be the most accessible though all arrangements should be checked by a calculation such as in equation {1} or by a scoring systems, as in Method 3 of MIL-HDBK-472 [1].

The influence of standardisation on maintainability hardly needs emphasising. However, although the number of different types of parts is often minimised for economics, the need to standardise on tools and test equipment can be forgotten. Of course, standard parts often lead to standard tools, especially in the case of fasteners but standardising on test equipment can be more difficult. Consideration should always be given to providing scaled measurements via the BITE; e.g. providing electrical analogs of pressure and temperature, thus avoiding the need for direct measurements except where the BITE itself is in doubt. If this practice is followed, care must be taken that the scaling corresponds to that of the instrument used so that the maintainer is not expected to perform feats of mental arithmetic; e.g. it is asking for trouble to provide a pressure measurement whereby 0-7 bar corresponds to 0-10 volts - especially if the expected reading is, say, 5.3 bar.

Design factor	Maintainability factor									
	a	b	c	d	e	f	g	h	i	j
Human factors	•	•	•	•	•	•	•	•	•	•
Safety of personnel		•	×	×	•		•	•	•	•
Physical effort		×	×	•	•	•	•	•	•	×
Mental effort	•	•	×	×	×		×	×	•	•
Assistance from operating personnel	×	•	•	•	•		•	•	•	•
Number of maint. personnel						•	×	•	•	•
Assistance from contractors		×	•	•	•	•	•	•	•	•

Table 9 : Human Factors

52

It must be remembered that almost all maintenance is a human activity and that the influence of human factors on maintainability, as shown in Table 9, is very strong. Safety of personnel must always be of primary importance and it will often be necessary to compromise other factors so that this can be assured. While great efforts are often made to ensure that the maintainer's physical abilities (which may range from those of the 5th percentile woman to the 95th percentile man) are not exceeded, less attention is paid to mental requirements. Maintenance activities often take place under conditions of extreme discomfort and, in cases where the maintainer's survival depends on his or her actions (e.g. in a submarine), of emotional stress. Under such conditions, the maintainer should not be expected to make complex calculations or logical decisions.

Design factor	Maintainability factor									
	a	b	c	d	e	f	g	h	i	j
Maintenance induced failures		•	×	•	•	•	•	•	•	•
Identification of parts & ass'ys		•	×	•	•	×	•	•	•	•
Captive fasteners						•	•	•	•	•	•
Incorrect replacement, part or ass'y				×			×	•	•	•
Need for preventive maintenance						•				•
Protection against secondary failure	•									•
Need for adjustments			×	×	•			•	•	•

Table 10 : Maintenance Induced Failures

Table 10 summarises the influence of design factors on the tendency for maintenance induced failures. The clear identification of all parts, modules, and sub-assemblies, the use of captive fasteners, the 'keying' of items to prevent inadvertent insertion in the wrong place, and the avoidance of the need for adjustments are all well recognised factors which can help to reduce the occurrence of such failures. Preventive maintenance should be avoided, if possible, on the grounds that if a task does not have to be performed then it cannot be performed incorrectly. The provision of devices to prevent secondary failures is less well known. In electronic systems it is (or should be) common practice to place diodes in series with the power supply lines to prevent the circuit's destruction if the supply is connected backwards. Similarly, all test points should be buffered to prevent damage by accidental short circuits. Barriers can be provided between electrical and mechanical parts of a system to prevent damage by heavy tools. Flexible pipes and cables can be arranged to enter equipment from above so as to minimise the possibility of damage by being trodden on. Unfortunately, there are few general rules and each case must be treated on its merits.

7 CONCLUSIONS

This paper has attempted to give no more than the barest outline of maintainability and its influence upon the integrity of safety-critical systems. An examination of the system type and its mission characteristics can indicate possible maintenance strategies and these can then be examined in the light of safety requirements for the individual tasks. The maintenance process can be reduced to a sequence of activities whose influence on both the integrity of the system and the safety of personnel can be examined in detail. Finally, the influence of design characteristics on the different maintainability factors has been discussed. For further details, the reader is referred to the various texts and papers listed below.

REFERENCES.

1   MIL-HDBK-472 *Maintainability Prediction*. U.S. Department of Defense.

2   BS 3811:1984 *Maintenance management terms in terotechnology*. BSI

3   Whetton, C.P. 1993 *Maintainability and its Application to Process Plant*. Process Safety Progress. Vol. 42, No. 3, pp 158-165

4   Pitblado, R.M., J. Williams, and D. Slater. 1990 *Quantitative Assessment of Process Safety Programs*. Plant/Operations Progress. Vol 9. No 3. pp 169-75

5   O'Shima, E. 1989 *Maintenance Practices in Japanese Process Industries*. Proc. IFAC Production Control in the Process Industry. Osaka, Japan, 1989. pp 189-94

6   Miwa, T. 1989 *Reduction of Breakdowns through Productive Maintenance*. Ibid. pp 199-207

7   Cornett, C.L. & J.L. Jones. 1970 *Reliability Revisited*. Chemical Engineering Progress. Vol 66. No 12. pp 29-33

8   Horrel, W. R. *Alignment: the engineer's responsibilities*. The Chemical Engineer, 17 January, 1991.

9   Kiernan, V. *Own up, says NASA, we won't sack you*. New Scientist, 7th August, 1993. p4

10   Anon. 1992 *Dangerous Maintenance*. London, HMSO. ISBN 0-11-886347-9

11   Whetton, C.P. 1992 *Thermohydraulic sneaks*. In Proc. Sneak Analysis Workshop. ESA-WPP033, European Space Agency, ESTEC, Noordwijk, The Netherlands.

12   Whetton, C.P. 1991 *Maintainability*. Sheffield University. Module 4 of MSc course in Process Safety and Loss Prevention.

13   Lyonnet, P. 1988 *La maintenance: mathematiques et methodes*. Paris, Techniques et Documentation (Lavoisier)

14   Blanks, H.S. 1992 *Reliability in Procurement and Use*. Chichester. John Wiley and Sons. ISBN 0-471-93488-7

SAFETY CRITICAL PROBLEMS IN MEDICAL SYSTEMS

Brian Davies
Mechanical Engineering Dept., Imperial College,
London SW7 2BX, England.

ABSTRACT

Advanced medical systems have many safety problems concerned with hardware, software and the human/computer interface. The paper discusses these as general issues and then goes on to illustrate them with particular reference to robotic systems in medicine, with which the author has direct experience. These robotic systems are discussed in some detail and illustrated with examples from robotic surgery and from rehabilitation robotics. As a result of safety issues that arose in this work, a series of suggestions are made in the hope of providing a start to a general consensus on what should be incorporated into an advanced medical safety system. Only when such a consensus is achieved can the most appropriate level of safety system be designed for a particular application.

1. INTRODUCTION

Medical Systems may involve activities which support investigation, diagnosis or treatment in any combination. Through incorrect design, malfunction, misinterpretation or negligence, the system may result in outcomes whose seriousness may range from merely inconvenient through to injury and death. Whilst there is a temptation to put most cost and effort into those aspects which involve dangerous procedures, the unforeseen circumstance can convert the apparently trivial into a dangerous outcome. Thus the classification of applications into those which could have a dangerous outcome (and which therefore warrant high cost and effort) must be carefully considered. Medical systems may be concerned with gathering data, with its interpretation to provide new knowledge and information, or with the subsequent diagnosis and treatment. The systems may combine aspects of human intervention and automation. The safety of medical systems may thus depend on a range of hardware, software and human/computer interfaces (HCI), any of which may be safety critical.

The author has particular experience of the use of robots in medicine [1]. This activity incorporates, in essence, most of the problem areas that can be met in complex medical systems. Not only are data gathered on the basis of which actions are taken, but these 'actions' are performed by a robotic system which involves prime movers (often of considerable power) being used next to people. These robots involve a computer controlled mechanism, dependent on safe software operating on the basis of imaging and data gathering systems which are themselves computer controlled. The human operators involved often have to interpret information and initiate subsequent actions that are carried out automatically by a mechanism. The concept of robot safety has been widely discussed [2] and is of particular importance where non-standard robots (known as 'Advanced' robots) are used in novel applications [3].

Medical robots utilise a computer control system that controls prime movers which can affect a patient's safety. This makes them a useful vehicle to explore the more general safety concepts for advanced medical systems. An example is the case of lithotripsy. Here a high power focused ultrasound beam is used to disintegrate stones in the renal system. Ultrasound images are taken which can give 3D coordinates of the stone via an operator interacting with a computer system. The coordinates for the target stone are then fed into a motorised system which moves the ultrasound transmitters so that they focus together at the target. The system thus relies on the integrity of software for the correct control of the motorised transmitters. The HCI is critical to ensure that the correct actions are taken in an environment which is prone to emergencies. Fortunately, the result of a malfunction is seldom life threatening, although the result can be very painful. The same cannot be said for the malfunction of advanced anaesthesia equipment where a computer controlled pumping system is life critical and is dependent on correct action of both software and hardware. The safety of such critical systems must be assured. Recent cases of the malfunction of radiotherapy equipment, in which incorrect gains were input manually into a program controlling dose levels, point to the importance of an unambiguous human/computer interface in which the operator input is clearly specified. In all these instances we see examples of the Engineer having built into the hardware or software, decision making tasks which were previously taken by medical personnel. Such decisions are often taken without the explicit knowledge and permission of the clinicians. The gradual change in responsibility for decision making, embodied within high technology systems in medicine, needs to be clearly discussed and the implications recognised by all concerned.

Robotic systems in medicine involve the whole range of problems that can be found in medical systems. They affect the safe operation of hardware, software and the human/computer interface and thus form a good basis for detailed study which will lead to high lighting many of the problems of safety in medical systems.

2. MEDICAL ROBOTS

Medical robots are examples of medical equipment and so must conform to the normal European standards affecting the use of medical equipment. A few of these have been recently formulated but the majority are still to be agreed by the Community. One problem is that whilst the requirements of simple components (such as plugs and sockets) can be readily defined, the design, for example, of programmable motorised control systems, is more difficult. In the past, National Standards have used such phrases as "shall conform to the best current practice" which has left the system developers in considerable doubt as to what is required and can only be tested by expensive litigation after a failure occurs. If the European Community standards are to be of benefit, then it is essential that they 'grasp the nettle' of safety specification in this difficult area.

A major problem is that the use of a powered robot next to people is generally necessary for it to be useful in medicine. This means that it is outside the Health and Safety Executive recommendations for industrial robots. These specify that the robot must be operated inside a cage from which all personnel are excluded. If people are present (for example, to program the robot) then the powered robot must be restricted in speed at that time. The two major uses of medical robots are in rehabilitation and in robotic surgery. In both of these instances it would be extremely limiting if the patient and clinicians were excluded from the reach of a robot and robots would cease to be a viable option. Thus it is necessary for a new set of guide lines to be generated in which the use of the robot is defined and the levels of safety implementations are specified. It will be necessary for a range of people to join in a debate and come to an informed agreement as to the requirements. This will require the Engineering and Computing Communities to decide what are a reasonable series of options available for the various tasks, with a clear description of the advantages, costs and risk potentials involved. Only when these are clearly presented in 'jargon free' terms can the medical community judge what are the acceptable range of solutions to the medical tasks they have specified. These solutions will then need the ratification of the specialist users groups representing the various patient interests, the manufacturers and system suppliers, the public at large and the Government legislative bodies.

One major problem is that increased levels of safety are generally achieved by increasing cost and complexity. How safe the system should be for different procedures will need careful discussion. It is generally recognised that even where safety is of over-riding importance, eg in nuclear power stations, there is no such thing as total safety and errors in software and failures of hardware do occur, in spite of duplication of systems and the very high costs which ensue. What is needed is a recognition that the benefits to be obtained from medical robots are such that a small amount of risk is inherent in their use and this is justifiable and acceptable. This is not to say that unsafe or unsound medical robot systems should be utilised. Every effort should be taken to ensure that the system is as safe as it possibly can be. Having done this, it is likely that some risk, no matter how small, will still be present. However, apart from some neurosurgery, the area of medical robots is almost never a case where the simple failure to perform its function will result in a life damaging situation. Provided that the medical robot is designed to fail in a safe manner and come to a controlled halt so that it can be removed and the procedure completed manually, there is in almost all cases no resulting danger to life . This is unlike the case of say, a military aircraft, which could not be flown manually if the computer control system failed. Medical robots can generally be removed and the procedure completed manually without any risk to the patient. Thus what is required is for the robot to come to a controlled safe stop in the event of a failure, rather than to have very long mean time between failures, which is a less important criterion.

Possibly the only exception to this general rule is in certain aspects of neurosurgery, where very critical regions are being operated upon. In such operations the exact trajectory of the tool and the accuracy of placement may be required to be so great that it is not possible to intervene manually without some detrimental outcome. This may be because the removal of the tool from the robotic system, and subsequent removal from the patient, is so difficult that it cannot be performed accurately and safely enough. Alternatively, in the case of critical locations deep within the brain, it is probable that the only way to achieve the required accuracies is to use a robot because the necessary accuracies could not be achieved with the use of stereotactic frames and jigs, (Subsequent to robot failure it is also unlikely that such jigs can be installed and positioned and then redatumed to the required accuracy). In such neurosurgical cases more emphasis will need to be placed upon using high values of reliability, because failure will be unacceptable whenever it is unlikely that there is a safe mode of failure that allows the device to be readily removed.

The question of how safe medical robots should be is not easy to quantify. If a medical robot allows a life saving operation to be performed, such as removal of a deep seated brain tumour, then one could argue that provided all reasonable safety measures have been taken so that failure of the system is rare, then the overall use of the robot was justified. Because its use would save lives that would otherwise be lost, the use of very expensive safety measures would also be easy to justify. However such arguments do not generally apply to the majority of robot surgery applications or to those of manipulators in the rehabilitation field. Here the robots are generally functioning as a replacement for human activity, simply because they are more accurate, faster, or do not require the continuing attendance of a person. The benefits that accrue are therefore generally less easy to quantify and are often concerned with cost saving or convenience. Since these benefits are not life critical, the safety measures needed and how much they can be justified in terms of complexity and cost are less clear. What is needed is for the medical community to make recommendations on the level of safety which is acceptable, with the various implications discussed so that all can come to a consensus of what is an agreed standard. If such a consensus is not achieved, then the development of medical robots will continue to be slow. Companies are understandably reluctant to develop new products where the required levels of safety with attendant legal issues are unclear.

3 Surgery Assistant Robots

As pointed out by the author in a previous paper [4], it is unlikely that it will be acceptable for surgery assistant robots to simply have a safer track record than human surgeons. Where failures occur, the robot supplier is likely to be held liable. Similarly, if the surgeon uses a relatively autonomous piece of computer controlled equipment which fails, there is less probability that patients will accept that surgery is risky and that the surgeon has done his best. Litigation is therefore more likely for robotic surgery than for conventional surgery. It is for this reason that the Author believes it is essential, when using a "robot assistant" in surgery, to involve the Surgeon wherever possible to confirm decisions throughout the surgical procedure. On the basis of the information displayed by the sensor systems and the human computer interface, it would then be the Surgeon's judgement whether to proceed with the operation. It is unlikely that this involvement of the Surgeon will totally deflect responsibility away from the equipment supplier. However, it should help to ensure that responsibility is shared.

Lavallee et. al. have adopted an aspect of Surgeon involvement for neurosurgery in France (5). Here the robot was used to carry a jig or fixture close to the head of the patient. When the fixture is at the correct position and orientation, a series of cutting instruments are clipped to the jig and used by the surgeon manually. Thus the robot is reduced to a preliminary role as a positioning jig and not for direct intervention. The safety of the robot is further assured by introducing a very large reduction gear ratio to each of the motor output drives. This means that in the event that something goes wrong with the robot, there is plenty of time to hit the emergency off button, because the robot is moving so slowly. This tactic however can only be successful when the robot is not in very close proximity to the head. Another suggestion is to use high level software to monitor the motions of a standard industrial robot. This may give rise to safety problems, because the monitoring is at a high level and not at the hardware servo level and hence may be slow to act. It is possible to imagine a scenario in which the servo is moving in a straight line when failure causes it to try to take off at high speed in the same direction. The inertia of the system may cause the robot arm to travel some distance before the fault is detected by the high level monitor and the brakes can bring the arm to a halt. It could be argued that only hardware monitoring at the servo level will give the necessary speed of response to avoid damage in critical instances, even though the robot is restricted to the role of carrying a jig or fixture. This modification will usually require information to be supplied by the robot manufacturers and thus needs their active co-operation.

A further aspect to this question is that it is seldom that the manufacturers of the industrial robot will make available to the suppliers of the medical robotic system, all the specific details of software and hardware in the industrial robot. Only if such full details are supplied can the medical group assure themselves that the necessary safety features are in place, so that they can take responsibility for the system that they are to supply. Another aspect to the use of industrial robots is that many industrial robot manufacturers specifically forbid the use of their systems in juxtaposition to people, since the devices were not designed with this application in mind and hence do not have the appropriate safety features. Given such a situation, it is the author's view that it would be foolhardy for researchers to use such robots unmodified in applications for rehabilitation or surgery. Whilst it may be acceptable to use a standard robot for feasibility study purposes to try out concepts and ideas prior to the implementation of a special purpose medical robot, care should be taken to protect the researchers. In the author's view, if the powered robot comes within range of the research worker, then similar safety features should be incorporated as would be used in the final device on a patient. Thus if, for example, a researcher in the laboratory holds the end of a 6 axis force sensor which is mounted on a powered up robot and "leads" the robot tip to a datuming position for simulated surgery on a cadaver, then that robot should have the same safety protection features for the researcher as if the application were by a surgeon for use on a patient. Whilst research applications do not require the same safety levels as for routine use, as evidenced by the relative ease of obtaining FDA approval in the USA for research, nevertheless, considerable care is still required.

Some redundancy in sensors and software systems is also desirable for reliability. Dual sensors, one on each servo motor and one on each output drive, could act as a check that the drive output system is not slipping and that encoder integrity is being maintained. Dual software systems running on separate transputers in parallel are another means of checking for errors. Additionally all software codes must be assured. Whilst dependable computing systems have made great strides, complex software still can not be guaranteed to be totally safe. One answer to these computing problems could be to have two software systems running simultaneously on two different pieces of hardware. For further integrity the software would need to have been independently generated by separate groups using completely different language systems. The two systems would need to be run in parallel and come to the same conclusion before any action could be taken. The cost and complexity of such a system would be likely to prohibit its use from all but the most life critical robot surgeon assistance.

One solution favoured by the author has been the design of a special purpose robot for a particular surgical procedure, to operate within a localised region. Associated with this concept is that of a mechanical constraint which physically prevents motion outside a limited area. This approach was arrived at because it was felt that the use of anthropomorphic robots capable of reaching a large volume space (the very feature which makes them attractive to industry) could, in the event of failure, fly off in any direction causing damage to patients, surgeons and other bystanders. A further implication of this approach is that, where possible, axes should be moved sequentially, one at a time rather than in parallel, to minimise the volume of space which can be swept out by an unforeseen motion so that appropriate protective measures can be adopted. This implies that complex shapes, resulting from the interaction of many axes of motion, may need to be approximated by a series of simple shapes, that result from a sequence of single axes motions. In addition to the usual sensor based software limits on motion, mechanical stops may need to be positioned for each programmed size of cut, to physically restrict the range of motion to a safe and acceptable limit should anything unforeseen occur. This concept of a special purpose robot with mechanical constraint has been further developed by the author in the design of a surgeon assistant for prostatectomies [6,7].

A unique example of a fully powered robot system for surgery is that being developed by Integrated Surgical Systems Ltd, USA. [8]. In Spring 1993, clinical trials were completed on the first ten patients for robotic hip surgery at Sacramento, USA. This system has been developed using a conventional industrial IBM robot whose control system has been substantially redesigned for the task with the aid of IBM. Additional duplicate position sensors have been used, together with a form of force sensor for each joint and much attention has been given to consideration of safety issues. However, the robot when standing on a base over 1m high, has an overall height of over 2m. The sense of power and size of such a system has caused some surgeons to express unease at having this robot operating next to them on their patients. Thus in spite of the considerable attention to safety, the perceived size and complexity of this system is likely to slow down its widespread application in the near future. in the longer term, this type of more generic system may find favour over the more specific, cheaper, simpler devices. However, the author believes that the latter will be the best approach in the near future

Many of the safety implications of robotic surgery are being avoided in the short term by the use of computer assisted surgery (CAS). A typical example is that at Aachen where ENT surgery is being carried out [9]. Here a passive arm is used to carry tools. Each joint on the passive arm is instrumented and, in some instances, can be locked using electromagnetic brakes. Since the surgeon is holding the tool and moves it directly to provide its power, he is in control of the process and relies on the computer system for tracking information. Thus the possible malfunction of robot arm servo systems is removed from the activity. The accuracy of the arm pointing system and the associated database can be checked out before hand, thus minimising problem areas to those of advising the surgeon rather than carrying out operations for him.

There is thus a host of tools potentially available to the surgeon, ranging from simple hand held tools, through various levels of 'Computer Assisted' Surgery, to the fully powered autonomous robot. An attempt has been made in Table 1 to classify these into a hierarchy of types in order to investigate the implications for safety. The table lists the type of system and gives some practical examples.

4 Robots in Rehabilitation

A further classification in which medically applied robots come into contact with people, is in their role as powered manipulators which assist the disabled in rehabilitation. Potential locations for the manipulator, relative to the patient, may range from no contact to full contact with the user. The former emulates the industrial set up, so that the medical manipulator is part of a workstation with an envelope of reach which is always out of contact with the user and public and thus potential safety problems are largely avoided. In this application the robot can only place items onto, eg a rotary table, from which the user can then take them. Unfortunately, although potentially a very safe solution, it is also most limited in its benefits to the user. Also, it will be necessary to fence off the manipulator arm, away from both helpers and the public.

61

An alternative solution is one in which only the extreme reach of the extended manipulator arm comes near the user. This is often used as part of a workstation in which the arm is used as a feeding aid, presenting food at its full reach for the user to move the head, eg to eat from a spoon. Although this is a fairly safe solution, it is seldom possible to rigidly fix the user location. Thus if, say, the user slumps down in a wheelchair, then the user will be within the envelope of arm motions. Similarly long tools held in the gripper could also strike the user leading to unsafe situations.

Another solution would be to limit the arm to a very low force and speed, not just by software or, eg force control, but by the intrinsic capabilities of the arm. This has some merit because the disabled user is usually content with a slow motion, as long as there is not a long delay before an observable motion starts. However the low force capability usually implies that the arm must be capable of lifting the heaviest design load at the longest reach. Thus when the arm is in a bent configuration with a light load, it will still be capable of exerting a significant force. Also, even if so lightly powered that the arm can be pushed away with the chin, the resulting force capability could still damage the eye if the gripper were carrying a sharp object.

A solution of considerable attraction is to place the arm totally within reach of the user, for instance on the arm of a wheelchair. However it can then potentially reach (and hence damage) not only the user but helpers and passers-by. This can be partially guarded against by designating 'no go' areas for the patient, eg operating at slowest speed when the arm configuration is near the head and totally preventing motion where the arm could strike the head. This however has the problem mentioned earlier that it is very difficult to so constrain the user that the "no go" areas are constant with reference to the chair. Whilst this objection could be overcome with additional sensing that can dynamically adapt the no-go areas to suit the users current posture, this implies considerable extra complexity and cost and does nothing for the public passing by, who can adopt any position and circumnavigate no-go concepts. For them, more sensing all over the arm would be needed to warn of an impending collision and stop the arm. All such emergency procedures will require elaborate safety status checks before the arm can be re-initialised. It can be seen that as further systems are added to take care of potentially unsafe situations, the cost and complexity rises until the system becomes unlikely to be used. At some level of complexity, it is likely that we must all accept that medical manipulators are potentially hazardous devices and reasonable care will be needed when they are used.

When the medical manipulator is part of a complex control system, eg also involving a powered wheelchair with powered adjustable seat position and an environmental controller, then more problems occur. The integrated control system needs to take account not only of the individual devices, but of the potential unsafe outcome of a combination of the devices, eg attempting to pass through a doorway at speed whilst the arm is extended, or trying to turn the chair abruptly whilst the arm is carrying a load at full reach.

These complex systems were investigated in a European "TIDE" project for mobile medical manipulator systems, in which the author was concerned with safety issues (10). The TIDE project specified an international bus structure for a collection of sub-systems, (powered wheelchairs, manipulator and environmental controller). It also highlighted the particular problems for medical manipulators in this role. For the wheelchair and other systems a 'Dead Mans Switch' (DMS) has been recommended whereby for any prime mover to continue to act, a switch has to be positively held in an 'on-position'. The moment that the DMS is released the power is removed from the local prime mover power relays. However, such a severe stop would be disadvantageous for the smooth control of a manipulator, so instead it is proposed for the manipulator to have a DMS linked to a zero velocity command, fed into the normal control system. This has the disadvantage that the DMS is now reliant upon the integrity of the control system, but is probably the best overall compromise.

This concept is further supplemented in the TIDE proposal by a speed cut-off directly at the prime movers if the 'zero velocity control' command fails to act within a certain time span. Safety integrity is further enhanced by the use of safety monitoring, both of the central control functions and at the sub-system power module levels. The safety monitors would remove power from different areas, depending on the fault, and report back to the central display unit. In addition to the continuous action of closing a DMS, there is also need for a large red button, prominently displayed, as a final emergency 'off' switch.

5. Safety Suggestions for Complex Medical Systems and Robots

The author's experiences, both with the TIDE project and with the clinical implementation of surgical robots, suggest that, where possible, the use of the following features could be of benefit in improving safety levels for medical systems:-

a)   Provide safety sensors, monitors and isolation systems, at a servo level where they can act rapidly as well as at a supervisor level where they can take account of system interactions.

b)   Use a Dead Man Switch (DMS) concept (ie a continuous positive signal is required to proceed) rather than just an emergency 'off' button. The latter requires a positive action to first locate the button and then activate it, whereas a DMS simply requires the cessation of a continuous action. In some surgeon robot systems there may be some merit in placing the DMS at the end of the robot so that the operator holds the robot tip and DMS whilst leading the robot into position. The ability to feel forces and sudden changes of acceleration/velocity may enable the user to have a faster reaction time in letting go of the DMS and stopping the robot.

c)   Use a safety monitor at both a servo level and a supervisory level to allow a rapid response to cut prime movers.

d)   Provide a facility to cut out prime movers independently whilst still keeping micro-processors powered up. Such a facility should be provided at a local (servo) level as well as at a systems level.

e)   Duplicate sensors. Monitor motions at the final resulting output as well as at the motor shaft to avoid unnoticed slipping of coupling and power transmissions.

f)   Avoid adapting software models/motions on-line. Try to check out all motions and models in a pre-process planning phase.

g)   Avoid the use of Artificial Intelligence (AI) systems except for preliminary planning phases to ensure maximum predictability.

h)   Arrange motor force capabilities to be inherently limited to just satisfy the required tasks, rather than only relying on force sensors and software systems to restrict forces.

i)   Use robust mechanical constraints to limit motions to a 'safe' volume/trajectory in case all other safety measures should fail.

j)   In the event of failure i) ensure the system can only fail in a safe, predictable manner, ii) ensure the system can be readily removed, iii) ensure that when the system is restarted, it has not moved relative to the patient (e.g. use absolute position measurement. Monitor the position of the patient on-line with reference to a robot datum). Alternatively ensure that when motors are off, the robot can be back driven by hand to a known start position to re datum the robot.

k)   Use a key switch and key line to power up the system, with full safety status checks at power up. Use a number of emergency "off" buttons, placed for easy access, rather than rely only on the key switch.

l)   Use single axis motions sequentially where possible, rather than compound axis motions, to limit the number of possible outcomes (range, motions, etc) should a failure occur.

m)   Give the user clear error messages indicating the current status of the system. Keep the HCI as simple as possible to avoid confusion but include all vital data.

n)   When datuming a robot to the patient for use in robot surgery, attempt duplication of datuming features used in the imaging process, eg the use of surface markers visible in CT scan together with a skin dye mark. This can give an independent check on the validity of the referencing system, rather than just rely on, for example, intra operative datuming of anatomical features.

6   Conclusions

As medical systems develop into areas of high technology, they are starting to incorporate many of the features which cause unease about safety integrity. They often have one, or all, of the aspects of on-line computer control systems which use software to control some type of prime mover. Software is also used for translating imaging data into 3D models, for overall management of the task and for communication of the clinicians desires to the control system. This combination of software, hardware and human/computer interface gives rise to many of the classic safety critical problems. The use of software with some type of action automatically resulting, is also giving rise to decisions embedded in software which have traditionally been taken by the clinicians. This process often takes place with Engineers talking in isolation to a particular group of Clinicians with the result that decision making is taken by the system without the wider group of Clinician users being either aware of, or being a party to, the discussions about the implications of such decisions. This can result in inappropriate and ill informed actions by Clinicians which can adversely affect the patient. Wider discussions on a framework for incorporating such safety related decisions into medical products is urgently required.

Software controlled mechanisms which require some type of human interaction can be found in medicine in areas as diverse as imaging (eg. for ultrasound systems as part of automated mammography or in Magnetic Resonance Imaging (MRI) systems for whole body scans); in 3D modelling (e.g. when building a picture of a brain tumour from a composite of different types of images); in Lithotripsy (where ultrasound pictures of stones in the renal system are identified by an operator to automatically focus high intensity ultrasound onto a target in 3D space); and in robotic surgery (where pre operative images are turned into 3D models. These are then used to target a robot to a patients anatomical features via intraoperative datuming). This last example has been considered in detail, together with rehabilitation robots, to give a picture of the typical problems encountered in advanced medical systems. These have given rise to a suggested hierarchy of robot surgery procedures which, as they become more complex, embody more engineering aspects that can fail. However, together with their increased complexity, these systems can take the control away from the medical personnel. This has the short term effect of reluctance by medical groups to relinquish control to the machines and resulted in concerns by equipment suppliers that they will be totally liable in the event of an operation failure. This has the effect of slowing down the pace of system development. However, such considerations do acknowledge the problems of human error if control is left too much in the hands of medical personnel. Then the high degree of concentration and judgement required inevitably leads to strain and incorrect actions. If, for example, high speed cutting tools are in the hands of a surgeon without any constraints then the result is likely to be less safe than a fully developed robotic system, in which a constraining mechanism is monitored by proven safety systems. Thus the longer term result of a fully developed complex medical system, in which the safety problems have been overcome, will be a safer system than a predominantly manual one.

A number of suggestions are given in the hope that they will help in the development of safer medical systems. However, wide ranging discussions are needed to achieve a consensus of what is necessary to provide adequately safe medical systems within cost and complexity constraints. Over prescription of safety features will prevent the development and application of medical systems which can save lives. Inadequate attention to safety will be just as damaging.

References

1.   Davies, B.L., Hibberd, R.D., "Experiences with Robots in Medicine" at Imperial College London".Proc. of IEEE Conf. on systems, man and cybernetics, Le Touquet, France, Oct. 1993.

2.   American National Standard for Industrial Robots and Robot Systems - Safety Requirements ANSI/RIA R 15.06 - 1986. American National Standards Institute, New York, 1986.

3.   Schofield, M. "Safety and Standards for Advanced Robots". Advanced Robotics Research Ltd., Salford, UK. 1992.

4   Davies, B.L., "Safety of Medical Robots". Chapman and Hall, Book Safety Critical Systems, Ch. 15, pg 193-201, 1993.

5.   Lavallee, S.," A new system for computer assisted neuro-surgery". IEEE Eng. in Med. & Biology Soc. 11 Int. Conf. pp 926,927, 1989.

6.   Davies, B.L., Hibberd, R.D., Ng, W.S., Timoney, A., Wickham, J.E.A., "Development of a Surgeon Robot for Prostatectomies", J. Eng. in Medicine, Proc. H. of IMechE. Vol 205, M.E.P. ltd., July 91.

7.   Davies, B.L., Hibberd, R.D., Ng, W.S., Timoney, A., Wickham, J.E.A., "A Surgeon Robot for Prostatectomies", Proc. 5 th I.C.A.R. Pisa, Italy; IEEE. U.S.A., June 91.

8.   Cain, P., Kazanides, P., Zuhars, J., Mittelstadt, B., Paul, H. "Safety Considerations in a Surgical Robot". Proc. ISA Conf; paper 93-035, 1993.

9.   Adams, J. Gilsbach J., Krybus K., Ebrecht D., Mosges R." CAS.- A Navigation Support for Surgery." IEEE Eng. in Med. & Biology Soc. 11 Int. Conf. 1989.

10   TIDE Project No 128 "A General Purpose Multiple Master Multiple Slave, Intelligent Interface for the Rehabilitation Environment.   E.C.E. DGX111/C3 Brussels, 1992

Table 1   Levels of complexity of systems in Surgery.

Code   Type of System

1.   Hand held Tools.
      Surgeon holds/moves tools freehand using only human innate sensing (touch, vision etc).

2.   Hand held Tools with a spatial location system.
      Freehand held tools, but surgeon can track a target using e.g. cameras + LED's on tool, or magnetic field source with sensors.

3.   Tools are mounted on a manipulator arm which is associated with a spatial location system.
      Arm is moved by surgeon, which to some extent constrains his sense of 'feel' and freedom of motion. Joint motions are usually monitored.

3.1   As for 3, but target location is updated with patient movement intra-operatively.
      Sometimes a 2nd (passive) arm is strapped to patient to monitor patient motion, on others a number of markers are tracked by an external camera system. Target location on a quantitative model is updated in real time to allow surgeon to track target with tool on arm.

4.   Tools are mounted on a manipulator arm which is associated with a spatial location system and powered brakes (passive).
      The addition of powered brakes to arm permits arm to be locked in position, eg to permit long term treatment at target location, or to permit surgeon to move away to safely fire x-rays.

4.1   As for 4 but the arm is used actively to insert/move tools.

4.2   As for 4 but the arm adapts to patient/organ movement intra-operatively.

5.   Tools are mounted on a powered robot arm equipped with position measurement (used passively).
Tools can be moved using the powered arm, either actively (as in 5.1) to enter the patient using the robot, or passively (as in 5). In the latter case the arm acts as a stationary jig which locates the tools so that the surgeon can manually insert them into the patient. In the active case, sometimes the robot is locked in position whilst a special purpose additional single axis moves into the body. In others it is the complete robot which moves to interact with the body. The ability to adapt, on-line, to patient motion (as in 5.2) risks the possibility of errors which cannot be trapped at a planning stage.

5.1   As for 5, but robot is used actively to insert/move tools.

5.2   As for 5, but robot adapts to patient/organ movement intra- operatively.

6.   Tools are mounted on a powered robot arm equipped with force and position control (used passively).
The addition of force control permits the robot to switch between position control and force control so that the robot can yield to a given force level in prescribed locations

6.1   As for 6 but robot is used actively to insert/move tools.

6.2   As for 6 but robot adapts to patient/organ movement intra- operatively.

7.   Tools are mounted on a powered robot arm equipped with force and position control with input systems from a "master" telemanipulator to control the powered robot as a 'slave' system. The use of data gloves and other tactile/force 'feed in' systems to the operator can enhance the operator knowledge of what is occurring at the 'slave' system.
Use of master/slave telemanipulator permits 'telepresence' with a data glove and 'virtual reality' to both control the slave and feed data back to the surgeon at the 'master' input. The addition of a master system as an input, additional to the normal computer control, increases complexity.

7.1   as for 7 but slave manipulator is used actively to insert/move tools

7.2   as for 7 but slave manipulator adapts to patient/organ movement intra-operatively.

Developing Safety Cases for Command and Control Systems

J.R Taylor
Taylor Associates ApS
Glumsoe, Denmark

1. Introduction

This paper describes a series of developments in techniques for the production of safety cases for Command and Control systems. The original objective was to be able to make full safety cases for military C^3 systems, but has been extended to a full range of computer controlled systems.

At the outset of the development (1986), quantitative criteria were established for C^3 systems, setting an upper limit on the frequency of large accidents with a potential for causing fatalities, of 10^{-6} per system per year. The use of a quantitative criterion of this kind has enabled decisions to be taken on a uniform basis. More importantly, the procedures necessary for risk calculation have been found to add considerably to the range of problems and failure possibilities identified.

2. Command and control systems and their problems

Command and control systems are widely used for military applications, air transport, space systems control, and ship control. Similar systems are used in nuclear power plants, chemical and petrol chemical plant. In modern practice they typically involve:

- some direct control capability for electronic and electromechanical hardware, including safety interlocking.

- a local area network.

- supervision and command software, controlling the operations carried out.

- monitoring and tracing software such as radar monitoring.

- communications software.

- man/machine interface software.

Typical C^3 systems involve from a few to tens of workstations, each with a few megabytes to tens of megabytes of software.

Obviously, such systems present a challenge to the safety analyst from the point of view of software reliability. Any system with more than a minimum of software will contain errors. Less obvious, is the fact that such systems generally also contain hardware design errors. Also, a major source of problems is the interaction between human error and weaknesses in the hardware and systems design.

By comparison with these design error problems, the problems of simple component failure are generally failure easily dealt with. The main design rules for C^3 systems and their accompanying hardware, namely fail safe design, redundancy, and interlocking, generally ensure that a high level of safety can be achieved, as far as hardwasre failure is concerned.

Because of the problems of design error related accidents, it is essential to apply an analysis strategy which accepts these problems. Because the problems often involve an interaction between hardware, software and operator, it is necessary to adopt techniques which can deal with all of these in an integrated way.

3. Limiting the Analysis Effort

A detailed analysis of a complete system with tens or hundreds of megabytes of software, and hundreds of circuit drawings, presents an almost impossible task. Even if the necessary effort can be made available, it is doubtful whether the necessary concentration, attention to detail, and overview could be mobilised for a sufficient period. The reliability of the analysis process itself then comes into question.

Two approaches have been used to obtain an effective analysis procedure.

The safety barrier analysis principle requires the analysis to start at the potential hazardous energy concentration e.g. rocket motor, igniter, firing mechanism. Causal paths are traced backward from this, moving through the active hardware, Control electronics, C^3 software, sensors operator actions, to the environment in which the C^3 systems works. At each stage, the safety barriers, which prevent accidents, are noted [6] (see fig. 1).

The safety barriers are assessed as follows:

- general quality of the barriers.

- possible barrier bypass routes

- possible barrier outage modes.

- reliability, including reliability of operation, and possibility of erroneous inactivation.

- possible common causes of failure with other barriers.

The assessment of barrier reliability proceeds along with the hazard identification and causal path route tracing. As soon as the reliability of the total set of barriers is found to be sufficiently high (e.g. unavailability $< 10^{-8}$), the analysis of the causal path is abandoned, or the level of analysis detail is much reduced.

This strategy allows the detailed analysis effort to be focused on hardware and on DDC software, with local area network software being analysed at the data flow diagram level.

The other strategy which has been used to limit effort is automation of hazard identification, as described below.

4. Satisfying a quantitative safety criterion for software

One of the main problems in developing safety causes for C^3 systems is the difficulty in analysing software. For limited amounts of software, it is possible to "prove correctness" of software modules, with respect to a specification. Practical limitations generally restrict the size of such modules to at most a few thousand statements. However, even when this is done, the "proof" is limited. A first problem is that the software specification may be discussed below. Secondly, there are many sorts of software interactions, such as time and resource competition, database interactions, compilation errors, which are beyond the current state of practice for correctness proofs, even in those cases where there are theoretical approaches available to deal with the problems.

A third group of problems arises from hardware failure interactions with software. A pattern sensitive or transient hardware failure can corrupt software or data, without becoming apparent. The results appear as software errors.

Some essential techniques for overcoming these kinds of problem are:

- restrictions of the safety critical parts of the software to an absolute minimum.

- isolation of safety critical parts, using either separate hardware, or segment protection hardware.

- use of fixed memory allocation, and simple multitasking priority schemes.

- use of array bound protection.

71

For the hardware failure/software interaction problems, a very simple technique is to use parallel redundancy, with comparison of outputs. This approach will detect most transient hardware failures, and many pattern sensitive failures.

Despite all these techniques, it is seldom possible to make a definite statement that a software will be free of hazardous failure. The best that can generally be achieved is a measured low overall failure rate (typically one failure per month, one failure per year, etc. which can be measured), and an estimated upper limit on the proportion of failures which are hazardous.

To determine an upper limit on the fraction of failures which are hazardous, a fairly simple technique has been used:

- The data flow paths through the software to the critical output registers are determined.

- The interlocks and consistency checks which are made by the software along the paths are listed.

- The interlocks are expressed in terms of Boolean criteria, and the domains of allowed outputs/not allowed outputs are determined.

- All failure are assumed to be potentially hazardous.

- The fraction of outputs able to cause accidents is the ration between the size of the allowed and not allowed output domains.

This combination techniques allows hazardous software failure rates of typically 10^{-2} to 10^{-3} per systems per year to be argued.

To argue a lower level of risk has required one of the following techniques:

- limitation of the opportunities for accidents (e.g. disconnecting hazardous subsystems until needed).

- additional safety interlocks, extraneous to the software.

- use of diverse redundancy.

5. Treatment of design error and specification error-Software FTA, Sneak Analysis, and Design Review

The problem of design error in C^3 systems are characteristically "system design errors". That is, errors which arise from activation of functions under wrong conditions, and failure to activate functions when required, or functioning to the wrong extent or degree, due to erroneous interaction between system components. This contracts with component selection and application errors, which are more seldom.

Functional analysis is a classic identification technique for system design errors. The technique involves considering all inputs and tracing their effects, under a rang of conditions. The difficulties with this approach are well known. The number of functional paths and possible conditions grows astronomically, even for small systems.

In hardware, sneak analysis provides a powerful alternative to functional analysis. In the variant which has been used, sneak path analysis, the tracing procedure starts by identifying "targets", that is items of hardware which can cause an accident if activated at the wrong time, or items of hardware which must operate in order to ensure safety.

The second step in the procedure is to find "sources" of inputs, that is power or signals, which can connect to the targets.

The third step is to find causal paths between sources and targets. These may be of two kinds, completely unintended paths, or intended paths which may be activated at the wrong time. Note that this kind of path search involves not only functional paths, but also, paths which have no intended function, but which are incidentally present in the physical construction.

The fourth step involves checking the conditions under which the given path may be closed. These "switching conditions" correspond to the operating modes, and possibly also some failure and error modes. A much more complete description of this procedure is given in [1]. Paths are characterised according to the number of failures and errors required for their activation.

For software, this kind of search is continued from the hardware, to the output registers of the computer, to the control flow in the software. Path tracing in software follows a simple back tracking algorithm, while the paths are traced, the "path predicates", that is, the input conditions allowing path to be followed, are collected.

If a path predicate becomes false, this implies that the path is impossible, and can be abandoned. One of the strengths of the method is that for a well designed system, there will only be a few possible paths.

Since this kind of tracing is pursued in a backtracking fashion. It is possible to record the results either as separate paths, or alternatively, in fault tree format.

The net result of this tracing is that system design errors involving unwanted activations, or unwanted cessation of a necessary function, can be identified systematically, and can be eliminated. This is true provided that the errors are apparent at the level of description analysed. In practice, the analysis will be pursued at high level such as data flow diagram level or functional block diagram level, and or at an intermediate level such as program source code. This means that the possibility of errors and failures at the lower level, such as compilation errors, need to be taken into account. They can conveniently be added to the fault tree, using check lists to identify the kinds of low level failures which can arise [2,3,4].

A feature of this approach is that it works with the actual design drawings and program code. It is to a very large extent independent of the system specification, and therefore also specification error.(fig. 2)

6. Automation of the approach

A fairly complete set of automated tools has been implemented in order to automate this approach, under the name MASER [5]. The functions provided (see fig. 3) include:

- functional analysis and functional failure analysis.

- hardware sneak analysis. [6]

- software fault tree analysis.

- human reliability analysis using the action error analysis method.

- computer aided design review.

These techniques may be used either separately, or in an integrated way.

The automation of analysis in this way has an obvious benefit in increased efficiency. The main benefit however is in terms of completeness and consistency of the analysis. Care has been taken in the implementation to ensure that all analysis steps are logged, and that the parts analysed can be distinguished from those remaining to be analysed.

7. Experience

The process, as described above, has been applied to three large C^3 systems, and a number of systems for chemical plant control. In all cases, the safety barrier approach allowed searches to be limited to just a few tens of hazardous signal paths - evidence of good safety design.

The kinds of problems which have typically been found in such systems have been of three kinds:

- the unavoidable low level errors and hardware failures, which presumably exist but cannot be identified. For these, an upper bounded on probability is determined.

- problems of unrecognised bypasses of barriers, such as transfer of signals from a system under test to a system.

- problems of planned signal paths, which can be activated at the wrong time due to operator error. These lead to identification of weaknesses in the man/machine interface.

8. Weaknesses and strengths of the approach

The strength of the approach described are:

- ability to deal with design error.

- ability to deal with most kinds of specification error.

- ability to deal with low level implementation errors probabilistically.

- efficiency, in that analysis effort is terminated in an area as soon as a sufficiently strong safety statement is achieved.

The result of the analysis is a statement of freedom from large classes of error, and a statement concerning an upper limit on the probability of remaining errors.

There are some weaknesses in the approach, which it is necessary to take into account in formulating a complete safety statement:

- there is a possibility that the actual system will not be built according to the design. This leads to a need to identify the possibility of construction and configuration control errors.

- some kinds of problems are parametric rather than logical i.e. they concern the size, rate, time delay of response rather than whether a function is present or not. While the techniques used can in principle be extended to cover these kinds of problem, the work in this area is still experimental.

- some of the judgements made concerning causality of event sequences are uncertain, e.g. whether a transient will be sufficiently large to trigger a threshold sensor. Such judgements need to be made conservatively. This leads to a need to make clear statements concerning assumptions, and the possibility that the assumptions will be invalid.

- some programs contain complex loops (this is not characteristic of safety related systems, but it does occur). In this case it may be difficult to determine a complete set of conditions under which a path may be followed. It may be necessary at time to redesign the system in order to guarantee analysability.

References

1. Bougnol C., Dore B., Taylor J.R. Development of Sneak Analysis procedures and Computer tools. ESREL conference, Munich 1993, Elsevier pub.

2. Phuh Westerheide, Quirk, Taylor and Voges, Software fault tree analysis in Verification and Validation of Real time software ed W.J.Quick Springer Verlag 1985.

3. NG Leveson, P.R.Harvey. Analysing Software Safety IEEE trans Software Engineering Vol BE-9 No5 Sept. 1983.

4. Taylor J.R. Fault tree and cause consequence analysis for Control Software, Risø-M-2326, Jan 1982, Risø National Laboratory, Denmark.

5. Taylor J.R. MASER, an integrated system for Hardware, Software and Operator Safety Analysis, ITSA 1991, Available from the author.

6. Taylor J.R. Risk Analysis for Process Plant, Pipelines, and Transport, Chapman & Hall 1993.

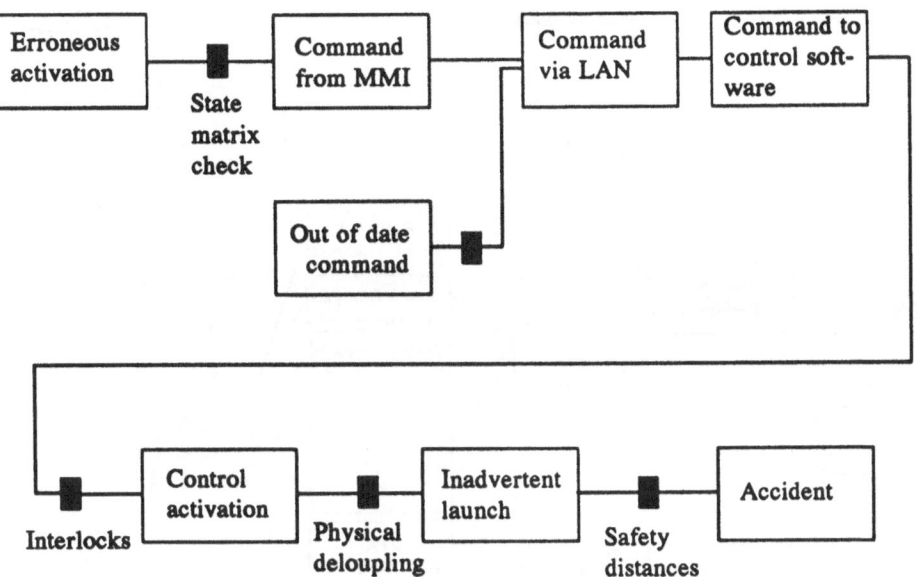

Fig. 1 Safety barrier diagram

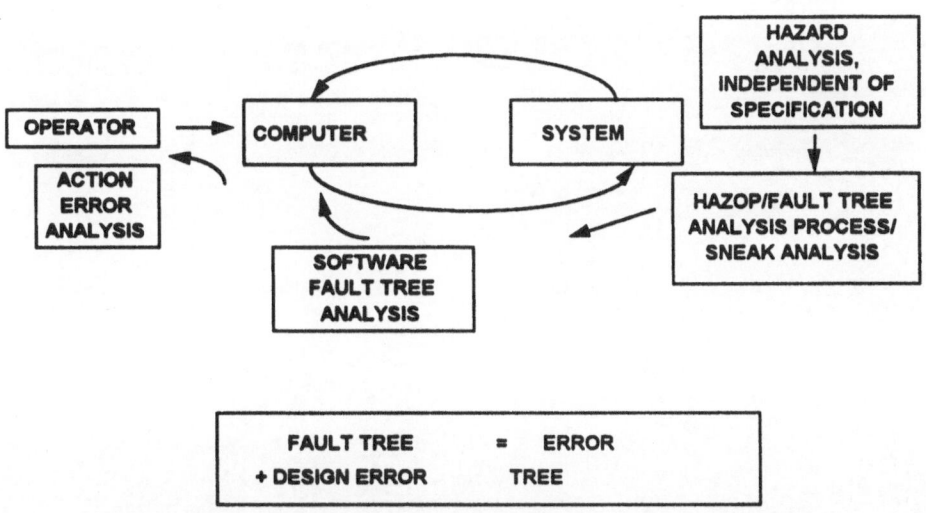

Fig.2 Integrated analysis for hardware/software/human systems

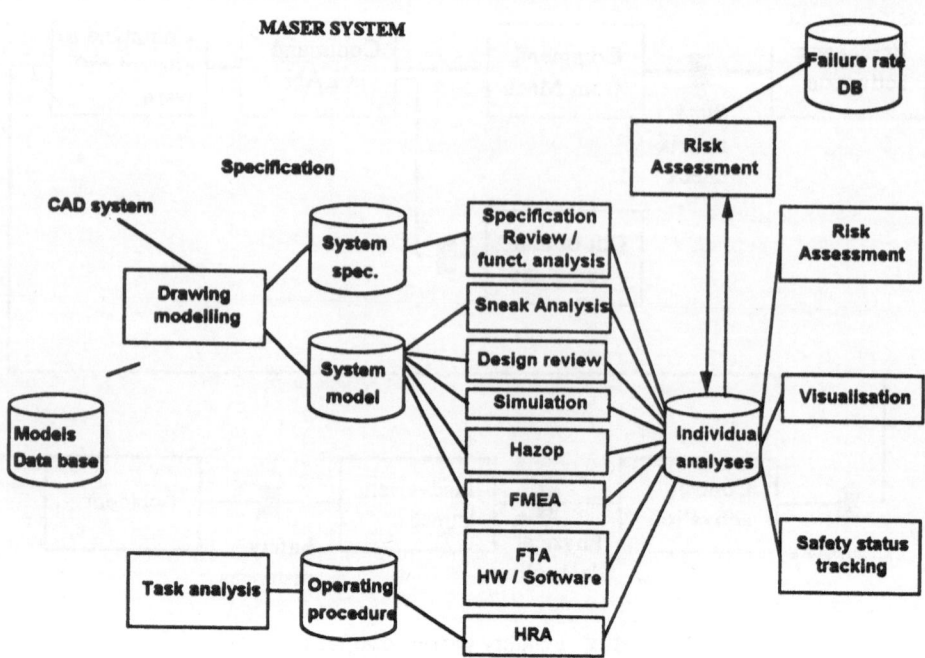

Fig 3. Maser system for C^3 systems analysis

Lifetrack: Organisational Modelling for Safety-Critical Decision Support

Author
Tony Holden, Michael Glykas
Decision Support Group, Engineering Department
University of Cambridge, Trumpington Street
Cambridge CB2 1PZ, UK.

Paul Wilhelmij   **Barrie Reynolds**
British Petroleum International   Honeywell Control Systems Ltd

Abstract

Petrochemical plants have inherent economic and personal risks associated with their operation. To give a concrete illustration of the scale of loss of life, personal injury and costs which can arise from plant incidents, the New York Times of 19th June 1991 lists 14 petrochemical disasters from 1987 to 1991 in the USA alone. Figures for these incidents were 7 dead, 138 injured and a cost of $246 million. It is therefore in the interests of the plant owner to apply best practices to ensure safe and cost effective plant operation, and it is in the public interest to encourage this.

There are two root causes of incidents leading to personal and economic loss. One is the lack of a trusted means for operational staff to access consistent, reliable and high quality information about the plant's overall operational status and the procedures for operating it. Secondly, there is the absence within management of a verifiable organisational model that accurately reflects the required behaviour of the organisational agents in performing safety-critical tasks. This is a necessary foundation to cope with the 'multifarious changes and unexpected events that occur in a complex plant.

The LIFETRACK project brings together an international petrochemical company (BP), a supplier of plant information and control systems (Honeywell) and an academic decision-support group. We are working together in a united attempt to make an in-depth investigation into the information needs of plant personnel when operating in safety-critical situations and how reliable information systems can be designed and maintained over the lifetime of a plant.

This paper illustrates how we have been developing the object-relationship modelling (ORM) and Object Life Cycle (OLC) languages for the problem of designing an organisational model as a prerequisite to an information system to support permit management. In addition to this, we also demonstrate the application of an object-oriented variant of the Z specification language to the verification of this model. Through this example, the principles of organisational modelling and verification for safety-critical tasks we are developing for the LIFETRACK project will be illustrated.

The information and opinions presented here are personal ones which do not reflect any official position or policy on the part of BP or Honeywell.

1. The drivers for safe and Profitable Plant Operation

Recent disasters involving petrochemical plants around the world have focused attention on the need to provide a higher degree of guarantee that plants are operating safely than has previously been required. Regulatory authorities are now requiring plant owners to move away from a prescriptive approach of merely requiring that a 'check-list' of predefined safety-critical requirements are met towards one that puts the onus of guarantee onto the owner to be able to demonstrate that the plant will be safe in all situations.

The high cost of litigation, coupled with these modified HSE (health, safety and environment) requirements are two significant business drivers influencing investment in well-designed decision-support systems. Another driver is the need to improve response times in increasingly aggressive market-places. Figure 1 summarises these influences.

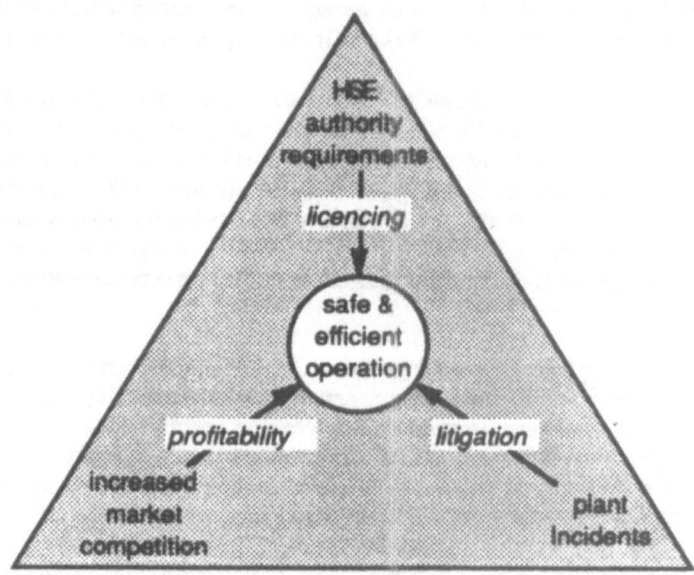

Figure 1: Business Drivers in Petrochemical Plant Operation

The foundation of safe and efficient operation is the availability of decision support systems for providing timely and accurate information on the state of a plant at any given time. This leads to improved operator understanding about the plant structure, operation and status. Understanding is the key to confident operation and hence safety (see figure 2).

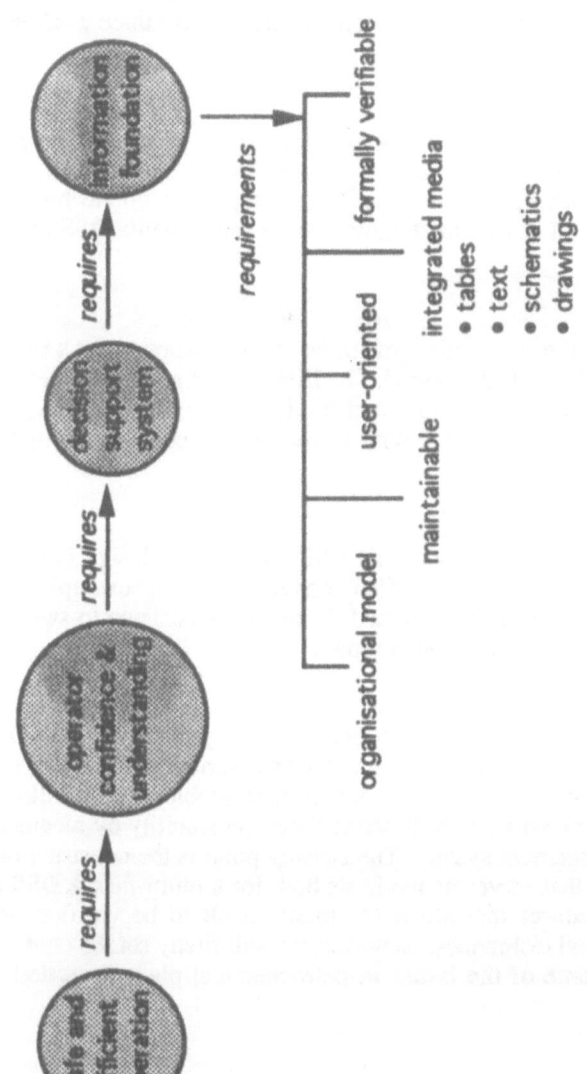

Figure 2: Foundational Requirements for a Reliable Decision Support System

However, existing decision support system design techniques have three weaknesses relevant to safety-critical decision support.

1)   They are designed to deliver predominantly table-oriented (eg SQL) information. In practice, engineers need to access other media such as tables, text, schematics and images. Although multi-media storage and presentation technologies exist, there are no established design methodologies for multi-media DSS as there are for table-oriented DSS.

2)   The need to *manage* the substantial quantities of information about a plant as it changes over its lifetime is rarely addressed. The lifetime of a plant from design to decommissioning can be 50 years or more and the information about changes in plant structure, operation, practices and lessons learnt must be consistently updated in a consistent fashion.

3)   It is difficult to be totally confident that DSS contains no design faults that could lead to a dangerous juxtaposition of different advice. A simple example of this would be the provision of advice that supported the mistaken decision to switch on a piece of equipment that was currently under repair.

Using our object-relationship modelling (ORM) and object-life cycle (OLC) techniques together with the Schuman-Pitt object-oriented variant of the Z language, we are attempting to provide a means for alleviating these problems. The following pages will explain these techniques and illustrate their applicability by means of a plant permit-to-work management system. The starting point is the construction of an 'organisational model' that serves as the basis both for a multi-media DSS and also for correctness procedures that allow the model itself to be verified using mathematically-based formal techniques. However, we will firstly set the context of the work by discussing some of the issues in petrochemical plant operation and information management.

1.1 Reasons for Choosing the Safety-Critical Domain

An advantage of selecting the safety-critical domain is not only driven by its obvious usefulness but also by the fact that the tolerance of people for their behaviour to be governed and constrained by 'formal' procedures is higher in safety-critical situations than it is for tasks where no similar health-threatening situation exists. In normal, day-to-day practice, people can find formal imposed approaches to job execution tiresome.

The notion of formalising other than a small circumscribed area of organisational task is also over-ambitious and so we are focusing on the problem of permit management in order to demonstrate our techniques. If these continue to be successful and well-received by those involved *in-situ*, we will consider expanding it to cover other tasks in the petrochemical situation.

1.2 The Permit Management Problem

A very significant area of expertise in the petrochemical industry that must be carefully managed is that of permit management. This the responsibility of the Operations Supervisor who applies his knowledge every time he takes on the responsibility of signing a permit to work on a live unit. The most tragic example of lack of safe permit management is the Piper Alpha disaster [Cul 91] where a decision was made to allow the operation of a pump that had its Pump Service Valve (PSV) removed for preventive maintenance, for which a permit was issued. The site of the PSV was not visible from the pump deck and personnel were apparently unaware that it had been removed. The consequences of this mistake were clearly high both in terms of loss of life and also financial revenue.

2. Sources of Risk in Plant Information Provision

Risks associated with plant information use, access and storage arise in four interrelated areas: with the *users* who require the information for operational decision making; within the *information* itself; within the information storage *architecture*, which, for reasons of practical access and maintainability, needs to contain information in an electronic form; and with *management* who has overall responsibility for unifying and co-ordinating the activities of the different and sometimes culturally diverse, interest groups (for example: designers, operators, maintenance workers, suppliers). A framework for hazard management must address all these areas.

The difficulty with providing accurate plant information arises out of the dynamic nature of petrochemical plants. Unlike other safety critical-systems, such as railway signalling networks for example, the process configuration changes in an on-going manner to accommodate evolving production and maintenance requirements, new operating procedures and enhanced process technology. Thus, over the lifetime of a plant there can be considerable structural modifications. As a result, ensuring that documentation is updated to reflect the current plant configuration is a non-trivial activity which has tended not be addressed successfully in practice.

2.1 Problems from the User Perspective

From the perspective of information users, inadequate access to plant information and procedures lead to operator confusion and uncertainty about necessary control actions and consequently to a lack of confidence in the control systems technology. This is particularly relevant when a response to an emergency is required. Poor access to plant status details can lead to plant personnel working at cross-purposes, thereby resulting in conflicting and possibly hazardous operational and maintenance activities.

Due to the inherent inflexibility of traditional documentation systems, documentation may be inaccurate with different versions containing inconsistent plant details. As a result, operational staff may not be well-informed about the current plant status, and different individuals may have conflicting views of the plant. This can lead to a lack of awareness of potentially hazardous developments or actions, and the consequent development of a hazardous situation that could have been avoided.

For continuous petrochemical process plants this is a particularly important issue during non-routine operations such as start-up and shut-down for both normal and upset conditions, equipment maintenance, and variable trimming in order to meet production targets. Experience has shown that the greatest risk to safety and profit arises during these times.

Improved access to information is also needed to help operational staff cope with the reduction in operational staffing levels, and with the increase in the complexity and level of automation of plant control systems.

2.2 Problems of Distributed Information

Information relating to critical decisions may reside in several dispersed locations, including the distributed control system (DCS), operations room, supervisors log and board and the engineers office. It may be in both paper and electronic form. Rapid access is therefore a problem, particularly as critical decisions may need to be made in a matter of seconds. Additionally, plant information should be generated and maintained to meet the needs of a range of plant users, and not the just one specific type of user, as tends to be the case at present. Additionally, in order to ensure that the plant information remains valid over the plant lifetime, it should be contained in a format independent of the storage technology.

Plant information is of a non-homogeneous nature, containing for example, textual documentation, process and instrumentation diagrams (P&IDs), and tabular process data. User-views which integrate non-homogeneous information and provide easy, quick access to information contained in different databases are needed, but are not provided by traditional software technology.

2.3 Problems of Information Management

Operating procedures and documentation which may be written before the plant is operational are often not adequately updated according to experience and plant modifications. Critical new information pertaining to revised procedures, control strategies and plant interconnections is often not reliably incorporated into the corpus of plant information as it is generated. For example, plant know-how generated during hazard and operability (HAZOP) analysis [Kletz 1986, CIA 1990], may not be passed on to operational staff.

84

It is therefore important to design flexibility into a DSS to ensure that changes to the plant can be accommodated and new types of operational users catered for. Standard interfaces should be used to allow additional databases and decision support tools to be added without complex integration problems. However, traditional software architectures are not designed to be inherently flexible or extendible, while databases tend to be designed for single types of user. Finally, for operational success, the software architecture and the information contained in it must be maintained by operational plant personnel, and not, as is presently the case, by information technology specialists.

2.4 Problems Arising from Cross-group Boundaries

A particular management issue is the need to ensure that the various engineering and technical groups that operate and maintain a petrochemical plant cooperate with each other and do not work at cross purposes. This can be difficult due to their distinctive technical and engineering cultures. A decision support system and associated information management framework will cut across both groups with different work cultures and values, and design and operational activities. These issues must therefore be considered.

3. Organisational Modelling Using ORML

To summarise then, different information is contained in different places for and by, different people. Designing a DSS with an underlying information architecture for supporting distributed, multi-user access is a difficult problem. Version control and consistency management are also non-trivial.

The next sections describe how organisational modelling and the object-relationship modelling language (ORML) may be used to aid in alleviating difficulties in these tasks.

3.1 Organisational Modelling as a Prerequisite for a Trusted Information System

In any management situation it is desirable to have a clear conception of what are the elements, or *agents*, that make up an organisation, and how they must work together in order to perform a task. Where these agents, whether they be people or technological systems, are co-operating on safety-critical tasks, this desire becomes a necessity.

More specifically, before a useful DSS of the type envisaged here can be produced, the following two prerequisites are needed:

• A methodology for analysing, designing and verifying a model of an organisation where users require heterogeneous information in order to perform safety- and cost- critical tasks.

• A management strategy for maintaining this information over the lifetime of the plant in a way that supports the belief that the decision-support system being used will always provide accurate information when required.

The starting point for an information architecture that meets these needs and is also demonstrably correct with respect to safety-critical needs is a clear organisational model. In LIFETRACK, we follow the philosophy of Ciborra [Ciborra 87] that organisations may be seen as *networks of transactions*, regulated over a period of relative stability by a *set of contracts* to govern relations between their members. The topic of organisational modelling through contracting agents is an involved one and outside the scope of this paper. Briefly though, organisations may be differentiated according to the nature of their constituent contracts. For example, we may distinguish between spot contracts in a market employment relation and authority contracts in beaurocracies, and long-term unwritten contracts within groups based on principles of mutual trust.

In Ciborra's view, contractors behave competitively as long as there is a conflict of interest. The use of information depends on the degree of policy congruence between the members of the organisation. If no policy congruence exists, information may also convey misleading signals and misrepresent goals, intentions and promises.

Organisational activities of co-ordination and control (which are of key importance in safety critical process management) may be understood using the notion of contracting processes. Four phases can be identified in the lifetime of a transaction:

1) A 'search' for parties involved in the exchange.
2) An identification of the 'terms' of the contract.
3) A control and 'enforcement' of contract execution.
4) The overall record and maintenance of the transaction.

Information is both needed and produced in the transaction activity and supports four different purposes:

1) It signals the willingness to bargain.
2) It elicits the terms of the contract.
3) It controls and enforces the contract.
4) It maintains communication during the transaction.

The information system of an organisation may be defined as the network of information resources required to conceive, set up, control and maintain the constituent contracts of the organisation. Thus each of the information systems required to support differing types of contractual processes has its own distinctive properties.

The more complex a given contract, the more difficult it is to achieve a communication system to regulate the exchange. Consequently more information has to be processed in order to set up and maintain the organisational relationship between the contracting parties.

We therefore see the information system as supporting the various processes necessary to promote and sustain organisational 'contracts'. In order for these contracts to be reliable and for the parties involved in them to work in unison and not at cross-purposes, a reliable information system must be based upon an articulated organisational model.

The ORML language has been developed to provide the formalism necessary for modelling organisations using contractual agents. The following sections describe ORML and its application to the permit management problem described earlier.

3.2 Object Relationship Modelling in Organisational Modelling for Permit Management

The basic constructs of ORM [Ip 91] are *agents* and *objects*. Agents represent the participants in contractual relationships and always include people. Objects on the other hand represent either transactions between agents or the resources that the agents utilise to discharge their responsibilities. A *relationship* is a general n-ary association between agents and objects. Relationships, objects and agents can be generalised to form classes which must have unique names. Each participating agent/object class in a relationship must have a responsibility. *Cardinality constraints* (min, max) have to be specified for both ends of a relationship. Figure 3 shows an ORM diagram that attempts to provide solutions to the permit management problem presented in the previous section. Objects and agents are represented by boxes and relationships between them by rhomboids. Binary relationships do not need to be given a name. The objects "Operator as Maintainer" and "Maintenance Management System" participate in the relationship class "Symptom Detection" through the responsibilities "reports" and "records" respectively. This relationship represents the fact that one operator is responsible for reporting 1 to N symptoms (he identifies) to the maintenance management system.

3.2.1 Complex Agents/Objects and their Responsibilities

Complex agents are the result of the aggregation of some other agents (which again may be complex) called participant agents. Similarly complex objects result from the aggregation of other objects called component objects. Double lined boxes represent complex objects and agents. For example in figure 3 the "Permit Management System" and "Maintenance Management System" are represented as complex objects. A special kind of object classes are value object classes . These can be either primitive system defined classes (integer, real etc.) or classes which only relate to other classes through attribute relationships. Value object classes are represented as boxes with a dark corner. The classes "Date", "Cost" etc. in figure 3 are value object classes of the class "Repair Work"

• Emergent Responsibilities

Some relationships between a complex agent/object and other agents/objects can only be meaningful if the complex agent/object is considered as a whole. The responsibility of the complex agent/object is then known as an emergent responsibility (abbreviated as E). The responsibility "issue" of a permit by the "Permit Management System in figure 3 is an emergent responsibility of the complex object.

- Delegated Responsibilities

A complex agent/object might participate in some relationships only because one of its participants or components participates in it. These responsibilities are called delegated (abbreviated as D). An emergent responsibility obviously cannot be delegated. The responsibility "advise" of the "Permit Management System" in figure 3 relates to the "Operations Supervisor" only because one of its component objects provides advise to him.

- Inherited Responsibilities

A component object can inherit some of its related objects (in particular, its attribute values) from its complex object. An inherited responsibility is marked with an I. By definition emergent responsibilities of a complex object cannot be inherited by its components.

- Contractual Relationships in ORM

A special kind of relationship that only holds between agents is called a contractual relationship . The aggregation of contracts in ORM presents a high similarity to the aggregation of agents/objects. The main difference between the two modes of aggregation is the definition of their boundary.

The boundary of an aggregation in ORM is defined as the set of all participants/components that have one or more responsibilities through which they participate in outside object/relationships. The direct components of a complex object must also be objects and the direct components of a relationship must also be relationships. Figure 4 represents the complex relationship "Symptom Detection". The relations related to the complex relationship by a pair of ispartof/haspart roles are direct components while all the others are called indirect. The relationship symptom detection consists of direct components only.

- Perspectives of Agents/Objects

A perspective of an agent/object can be seen as a special form of participant agent or component object. It is normally named as A-as-P where A is the name of the object class (known as the master class) and P the context that the perspective is applied. An instance of the agent/object class can have any number of instances of the same perspective but a perspective instance must, by definition, belong to one and only one agent or object instance (i.e. it is specifically dependent on the object). The problem of multiple responsibilities undertaken by people with the same job classification is overcome through the use of perspectives. For example in figure 5 we can see the four perspectives of the operator only one of which ("Operator As Maintainer") is relevant to the context of our example.

3.3 The Object Life Cycle

ORM is the underlying data modelling technique. An OLC [Ip 92b] can be constructed for every object class defined in ORM. Every object in a class must be in a certain *state* . The possible states of an object class are shown as circles in an OLC. An object can be transformed from one state to another by an *operation*. The input and the output states of an operation can be the same. If they are the same a double arrowed link is used. The notion of OLCs is applicable for describing the behaviour of resources in the organisation (see figures 6 and 8). The maintenance object class in figure 3 has its own OLC presented in figure 6. It contains three operations ("Safety", "Plant Monitor" and "Repair Work") and only one state ("Maintaining"). Organisational agents are modelled in terms of Agent Lifecycles (ALCs). The ALCs are modelled in the same way as OLCs but the concept of a state is substituted by the concept of an agent's role . Also, the ALC operations represent the actions through which the agent discharges its obligations to other agents (through access to relevant resources) by maintaining or changing the state of affairs. The ALC of planner is represented in figure 7. The planner undertakes two roles, namely: "Planning" and "Scheduling". When the planner undertakes anyone of these roles he is able to perform a different set of operations.

4. The Schuman-Pitt Object-Oriented Z Variant

The Schuman-Pitt methodology [Schuman 87, Schuman 90] is an Object Oriented variant of Z that uses a rigorous mathematically based notation for supporting the early stages of software design. It devotes special attention to the well established principle of data abstraction, which implies some basis for modular decomposition, into separately specified sub-units.

Structurally, the specification of an object class comprises two distinct kinds of definition: a state schema and an operation schema. A state schema in its general form is represented in figure 9(a).

The header of the state schema includes the class name (C) and the formal parameters a,b for the class. The declarations (e.g Xi:Si) introduce the state components (Xi) and their types (Si). The state invariants represent any constraints on the values of the state components and the initialisation conditions represent any additional constraints to be satisfied on initialisation.

A class specification also includes a number of self standing "operation-schemata", which define the various access operations that may be applied to an individual object to query and/or update its internal state. An operation schema is depicted in figure 9(b).

The operation name is E and the parameters a,b may be used in invoking the operation whereas the parameters c,d may be used for the values returned by the operation. The operations are characterised in terms of preconditions (above the double line) postconditions (below); the usual convention is adopted that dashed variables are used to represent the state component values after each occurrence of the operation. The preconditions specify what must hold for the operation to be applicable to an object of the class, whereas the postconditions specify the explicit effect of such an application.

Class composition (the term composition instead of aggregation will be used in the Schuman-Pitt schemata in order to distinguish from ORM aggregation) is depicted in figure 10(a). Class A is composed of classes B and C and their component classes.

A corresponding composition mechanism exists for the definition of operations. Operations can simply be "promoted" to the composite class and their associated constraints are taken to be inherited. Additional constraints and preconditions or postconditions can be added as necessary.

In figure 10(b) the composite operation A.a is represented in terms of operations B.a and C.a of its component classes with the additional preconditions Aprec and postconditions Apost.

5. Transformation from Informal to Formal Representation

5.1 Formalising ORM

For every agent and object in ORM a Schuman-Pitt class schema describing its state components, invariants and initialisation is constructed. One of the components of the schema is a set representing agent or object references. This means that an agent or object may evolve without changing the value of the reference agent or object in the set. The main reason for utilising this concept is that it allows the modelling of aggregates. This set is used to represent the ORM objects in all the relations they participate (relations between objects are not permitted in formal notations). The object class schema also includes the set of OLC states of the ORM object it represents and a function called 'ClassName_state' that specifies the state of a referenced object defined as:

ClassName_state : reference_objects -> set_of_states

The agent class schema includes the set of ALC roles and a function called 'AgentName_role' that specifies the role of a referenced agent defined as:

AgentName_role : reference_agents -> set_of_roles

These functions are particularly useful for the synchronisation of operations as we will explain later. Figure 11 depicts the Schuman-Pitt class schemata for the planner agent class and its components.

Responsibilities of agents and objects in ORM are represented as relations of their Schuman-Pitt class. The responsibilities of two participating objects in an ORM relationship are inverse relationships in the composite Schuman-Pitt class schemata. The class planner in figure 11 has as component the set of its agent references 'pl'. All the responsibilities of planner in figure 3 are represented as relations in the Planner Schuman-Pitt class schema in figure 11.

Cardinality constraints are expressed as invariants over the domain and the codomain of schema relations representing the ORM responsibilities. For a more detailed explanation of the representation of cardinality constraints in Z the reader should refer to [Polack 92].

Both inheritance relationships and the relationships between a complex object and its component objects are represented using class composition. Classes related to component classes of a complex class are included in the component class schemata. For example the class "Planner" in figure 11 includes the indirect classes "Repair_Work" and "Equipment" that is related to in figure 3 (the class Employee is a superclass of the class Planner). Both derived and emergent roles of a complex ORM object are represented as relations of the composite Schuman-Pitt class that represents the aggregate ORM object that includes the complex object as one of its components. ORM value object classes are included in the Schuman-Pitt class schema of the object that they relate to. The classes "Date", "Cost "etc. in figure 11 are included in the class "Repair_Work" and their attribute relationships are represented as components of the class

5.2 Formalising OLCs and ALCs: Towards verifiable safety critical processes

The operations of ALCs and OLCs are represented as operation schemata in Schuman-Pitt. Figure 12 describes the operation schemata for the operation "Repair of the "Maintenance" OLC in figure 4.

Of special importance is the notion of composition of operations. An operation in the composite class may be implemented in terms of operations of its component classes. For example, in figure 12 the operation "Maintenance.Repair" is implemented in terms of dix operations of its component classes. Five from the planner agent ("Estimate", "Resource_Identify", "Schedule", "Propose" and "Refer") and one from the maintenance management system object in figure 8 (Daily Scheduled Work).

The problem of ordering of operations of component classes that are included in the operations of their composite classes can either be left unspecified or a specific ordering can be presented. When several operations of a single component class are included in a composite class the ordering of operations can (usually) be deduced from the preconditions of the operations of the component classes. By examining the preconditions and postconditions of the four operations of the planner agent in figure 12 we can see that the operation "Planner.Resource_Identify" should be performed first. This is because the postcondition "rw_resource'(w) = r" is a precondition for every other operation of planner. The next operation to be performed is "Planner.Refer" since its postcondition "e_manual'(equipment $\wedge$ r) = e" (the operation ^l is the *image* of one set onto another) is a precondition of the remaining three operations. The next operation to be performed is "Planner.Propose" since its postcondition "rw_method'(w) = m" is a precondition of the remaining two operations.

As can be seen in figure 7 these three operations do not modify the role of the planner ALC. The precondition "Planner_role(p) = planning" is included as a precondition in all three schemes and the value of "Planner_role(p)" remains unchanged at the end of the operations leaving planner at the same role. The remaining two operations "Planner.Estimate" and "Planner.Schedule" can be performed in any order. The last operation, however, will change the role of planner to "scheduling" by having the postcondition "Planer_role'(p) = scheduling". This will only happen when the preconditions of boxes 1 in both schemes, that justify that the other operation has been performed, are satisfied.

When operations of more than one component class are included in a composite class schema the synchronisation of ALCs and OLCs is again achieved using appropriate preconditions that allow an operation to be performed if and only if the other agent or object is at a specific role or state. For example the operation "Maintenance_Management_System. Daily_Scheduled_Work" in figure 12 will be performed after all the operations of planner are finished since it includes preconditions like "rw_resource(w) = r" that belong to the postconditions of planner's operations.

In cases where the ordering of operations can not be determined all possible sequences of operations should be considered. In these cases the duration of a composite class operation is longer than the duration of any of the operations of its component classes that it includes. The composite class operation is active until all of the component class operations are performed.

5.2.1 Verification

It is widely recognised that strict adherence to formalised procedures such as the one modelled above can contribute to the reduction in process plant incidents. The advantage of applying the ORML methodology is that informal and formal models are made available for the analysis of more complex situations such as the Piper Alpha where more than one item of work was undertaken under a single permit to work. Moreover additional problems related to maintenance management, permit management, lessons learned etc. can all be resolved in a constant way using a single model of the overall process. This organisational model provides a clear high level view of the different activities within the enterprise and the way that they interact. It also indicates what information is required to permit the organisation subsections to exist and function together. Because it is a formal model it can be implemented using computer-based technology.

The transition from informal to formal representation in IS analysis is currently a very active research area. Many approaches try to provide a formal semantics for informal notations, such as dataflow, entity-relation or object-oriented diagrams, so that the formal model might be generated from the informal. Others prefer to use the structure in the informal model to decompose the problem. The resulting pieces are defined in a formalism appropriate for the class of problem originally conceived by the analyst, then recomposed to construct the formal model. This latter strategy we adopt in our methodology [Glykas 93b].

The use of Schuman-Pitt in ORML provides the option of verification of IS designs. The verification is achieved by proving that the model itself is logically consistent (that is, the model does not permit the derivation of mutually contradictory theorems). In the Schuman-Pitt methodology internal consistency of models expressed in it requires the discharge of the following proof obligations [Cohen 92]:

1.   *State consistency.* For every schema, it must be shown that there is at least one assignment of values to the state components which satisfy their type definitions and their invariant predicates.

2.   *Event applicability.* For every operation, it must be shown that there exists at least one state consistent assignment of values to the operation's schema, and one type consistent assignment to the operation's parameters which together satisfy the operation's precondition.

3.   *Event effectiveness.* For every operation, it must be shown that, for every consistent state and parameter setting in which the operation's precondition holds, its postcondition must satisfy the state invariant. (This must also be proved for the "initialisation" of each schema, which may be thought as an anonymous operation, with no precondition, which occurs when a system satisfying the schema comes into existence.)

The organisational processes in our framework are represented as sequences of Schuman-Pitt operations. By proving event effectiveness for every Schuman-Pitt operation we result having verifiable organisational processes [Glykas 93a] that are of particular importance in safety critical applications like permit management. The formal representation of these proof obligations as first order predicates is mechanically derivable from the text of the model itself.

6.   Implementation Status and Conclusions

The provision of reliable, multi-faceted information to operators and other users in petrochemical plants is becoming more important as the legal and economic demands of plant ownership become tighter. Operators must have a better understanding of, and hence, greater confidence in, the decisions they are taking.

A prerequisite to a stable decision support system in which plant operators and other personnel can have confidence is a formalised organisational model. By following the agent/contract view of organisational behaviour, we have developed the Object-Relationship Modelling Language (ORML) to provide such a formalism. If this DSS is going to support safety-critical operations, then it must be checked for validity. We have adopted the Schuman-Pitt Object-Oriented Z variant to do this on top of the ORML model.

Initially, we are focusing on the particular problem of permit management. Once we have refined our techniques on this area, we will then expand our horizons to cover a larger segment of petrochemical plant decision support.

The theoretical aspects of ORML are fairly advanced and have been applied to the real problem of permit management, we are constructing computer-based aids to help in the construction of ORML models. Ongoing work will continue with this and the development of automated verification aids. Figure 13 shows a screen dump from our graphical editor.

7. References

[Ciborra 87] In: Information Analysis: selected readings. Addision Wesley, Sydney. 1987.

[Cohen 92] Cohen B. The cbm company: An exercise in the formal specification of a dataflow analysis. In Proceedings of 5th International Conference: Putting into Practice Methods and Tools for Information System Design. Nantes, France, 1992.

[Cul 91] Lord Cullen. The Public Inquiry into the Piper Alpha Disaster. HMSO, London, 1991.

[Glykas 93a] Glykas M, Wilhelmij P, Holden T. Object Orientation in Enterprise Modelling. Proceedings of IEE Colloquium on Object Oriented Development. January, 1993.

[Glykas 93b]Glykas M, Wilhelmij P, Holden T. Verifiable Object Oriented Designs. To appear in Technology of Object Oriented Languages and Systems (TOOLS) 93 conference proceedings. August 1993.

[Ip 91]Ip S, Cheung L, and Holden T. Complex objects in knowledge based requirement engineering. Proceedings of the 6th Knowledge-Based Software Engineering Conference, September, 1991.

[Ip 92a]Ip S, A Knowledge Representation Approach to Information Systems Analysis and Modelling. Ph.D Thesis, Cambridge University, Engineering Department, 1992.

[Ip 92b] Ip S, and Holden T, A knowledge based technique for the process modelling of information systems: The object lifecycle diagram. In P.~Loucopoulos, editor, Proceedings of the 4th Conference in Advanced Information Systems Engineering. Springer-Verlag, Manchester UK, May 1992.

[NT 91] The New York Times, 19th June 1991.

[Polack 92] Polack F, Integrating formal notations and systems analysis: using entity relationship diagrams. Software Engineering Journal, 7(5), 1992.

[Schuman 87]Schuman S A, and Pitt D H. Object oriented subsystem specification. In L.Meertens, editor, Program specification and transformation. North-Holland, 1987.

[Schuman 90] Schuman S A, Pitt D H, and Byers P J. Object oriented process specification. In C.Rattray, editor, Specification and verification of concurrent systems. Workshops in Computing, Striling, Springer- , 1990.

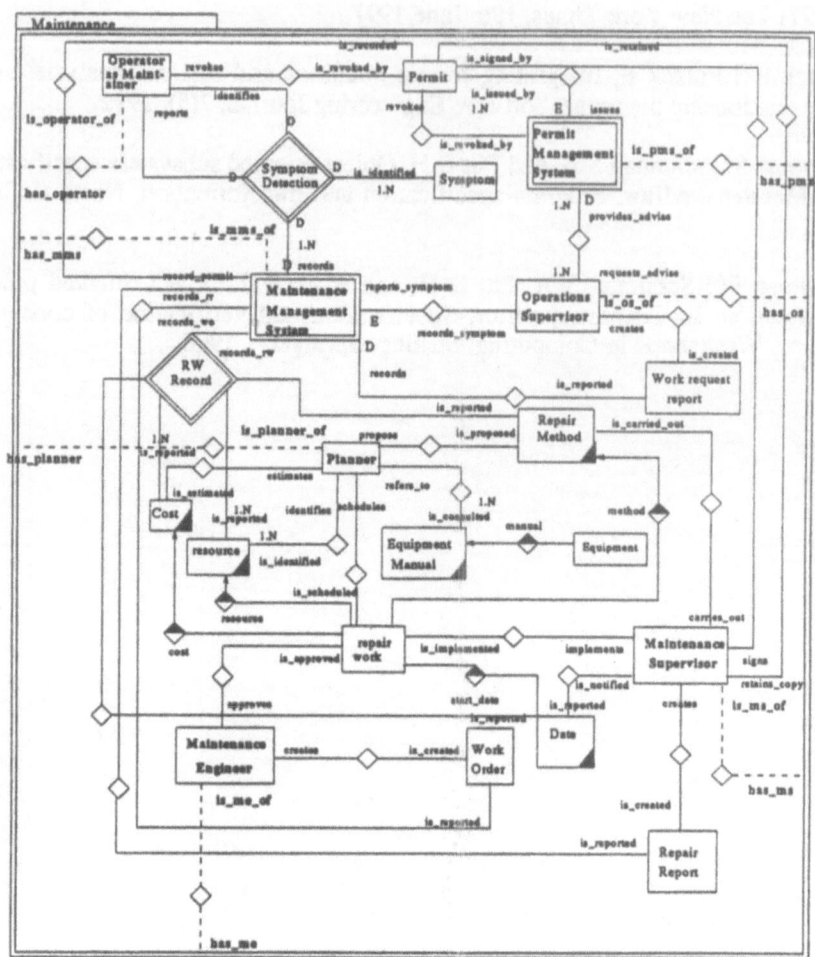

Figure 3. Plant Maintenance presented as a complex object in ORM

96

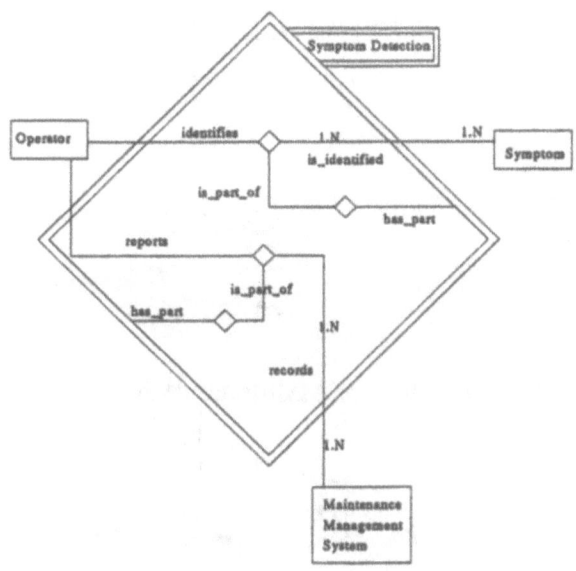

Figure 4. A complex relationship in ORM

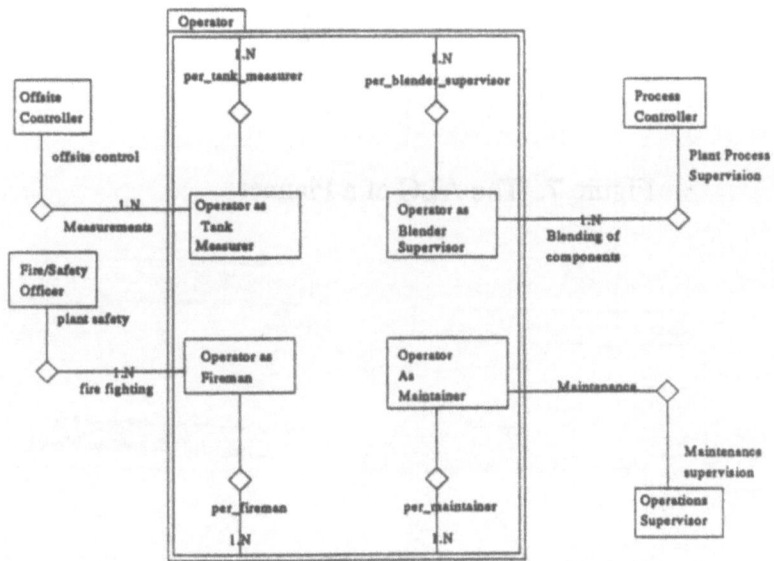

Figure 5. Organisation View in ORM

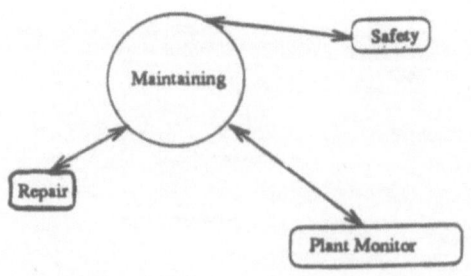

Figure 6. The OLC of Maintenance

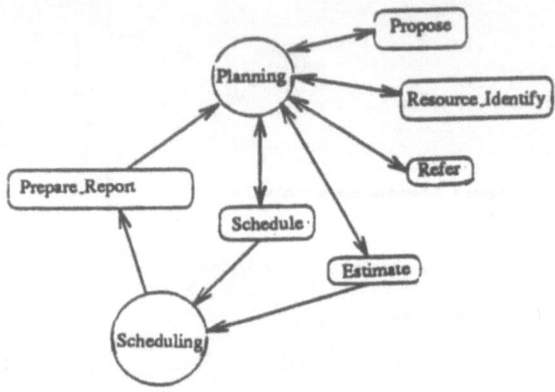

Figure 7. The ALC of a Planner

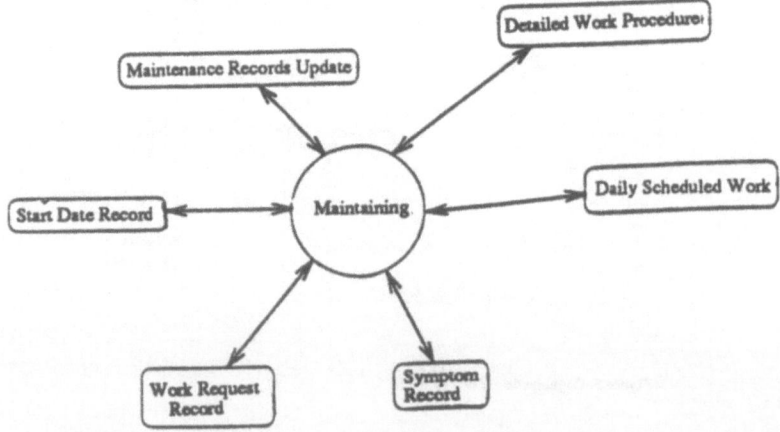

Figure 8. The OLC of the Maintenance Management System

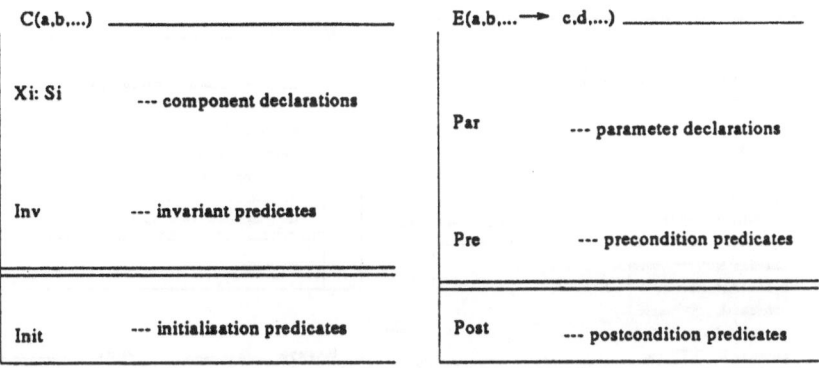

Figure 9 a. A Schuman-Pitt state schema

Figure 9 b. A Schuman-Pitt operation schema

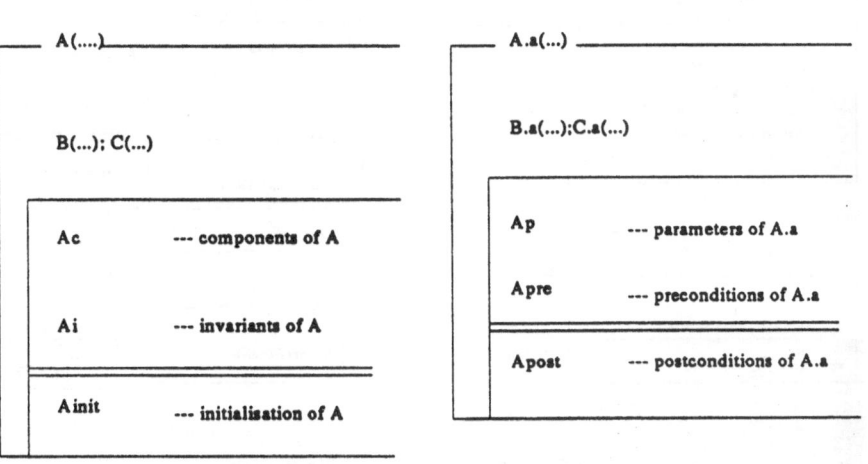

Figure 10 a. Class Composition

Figure 10 b. Composition of Operations

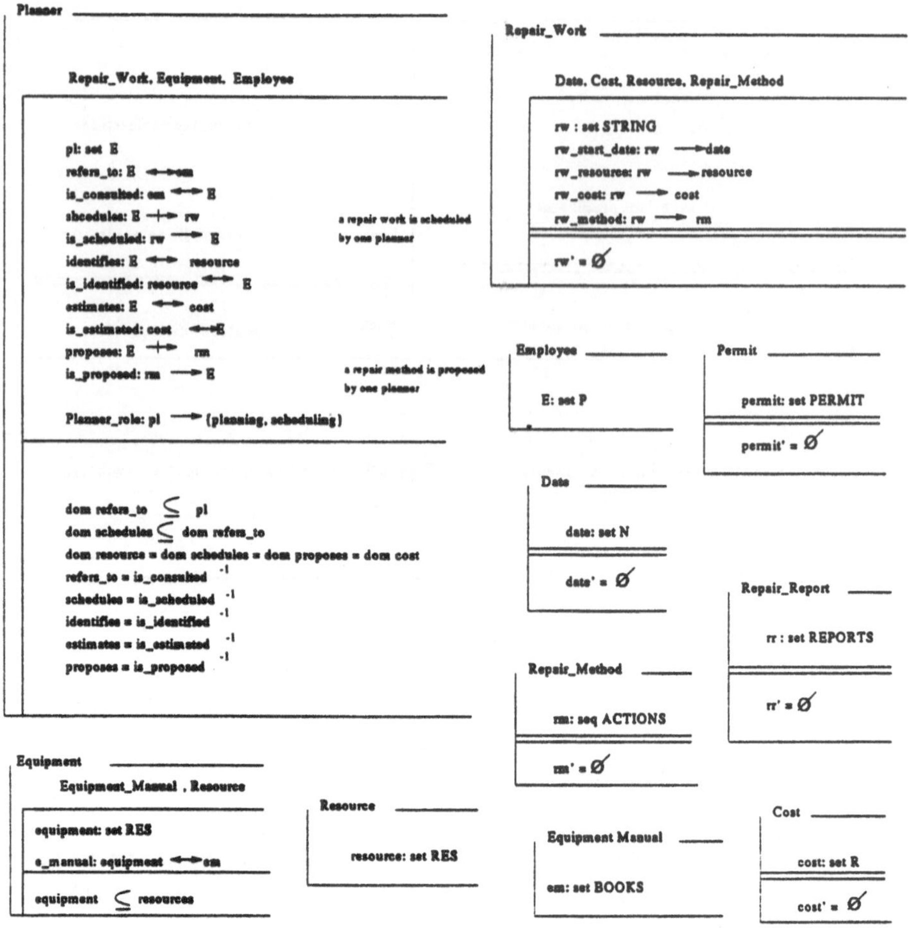

Planner —————————————

      Repair_Work, Equipment, Employee
—————————————————————

pl: set E

refers_to: E ◄—►em

is_consulted: em ◄—► E

shcedules: E —↦ rw   a repair work is scheduled

is_scheduled: rw —► E   by one planner

identifies: E ◄—► resource

is_identified: resource ◄—► E

estimates: E ◄—► cost

is_estimated: cost ◄—►E

proposes: E —↦ rm

is_proposed: rm —► E   a repair method is proposed

                               by one planner

Planner_role: pl —► (planning, scheduling)

dom refers_to $\subseteq$ pl

dom schedules $\subseteq$ dom refers_to

dom resource = dom schedules = dom proposes = dom cost

refers_to = is_consulted $^{-1}$

schedules = is_scheduled $^{-1}$

identifies = is_identified $^{-1}$

estimates = is_estimated $^{-1}$

proposes = is_proposed $^{-1}$

Repair_Work ———————————

      Date, Cost, Resource, Repair_Method
——————————————————

rw : set STRING

rw_start_date: rw —►date

rw_resource: rw —► resource

rw_cost: rw —► cost

rw_method: rw —► rm

rw' = ∅

Employee ——————

E: set P

Permit ——————

permit: set PERMIT

permit' = ∅

Date ——————

date: set N

date' = ∅

Repair_Method ——————

rm: seq ACTIONS

rm' = ∅

Repair_Report ——————

rr : set REPORTS

rr' = ∅

Equipment ———————————

    Equipment_Manual , Resource

equipment: set RES

e_manual: equipment ◄—►em

equipment $\subseteq$ resources

Resource ——————

resource: set RES

Equipment Manual ——————

em: set BOOKS

Cost ——————

cost: set R

cost' = ∅

Figure 11. Class Schemata for Planner and its components

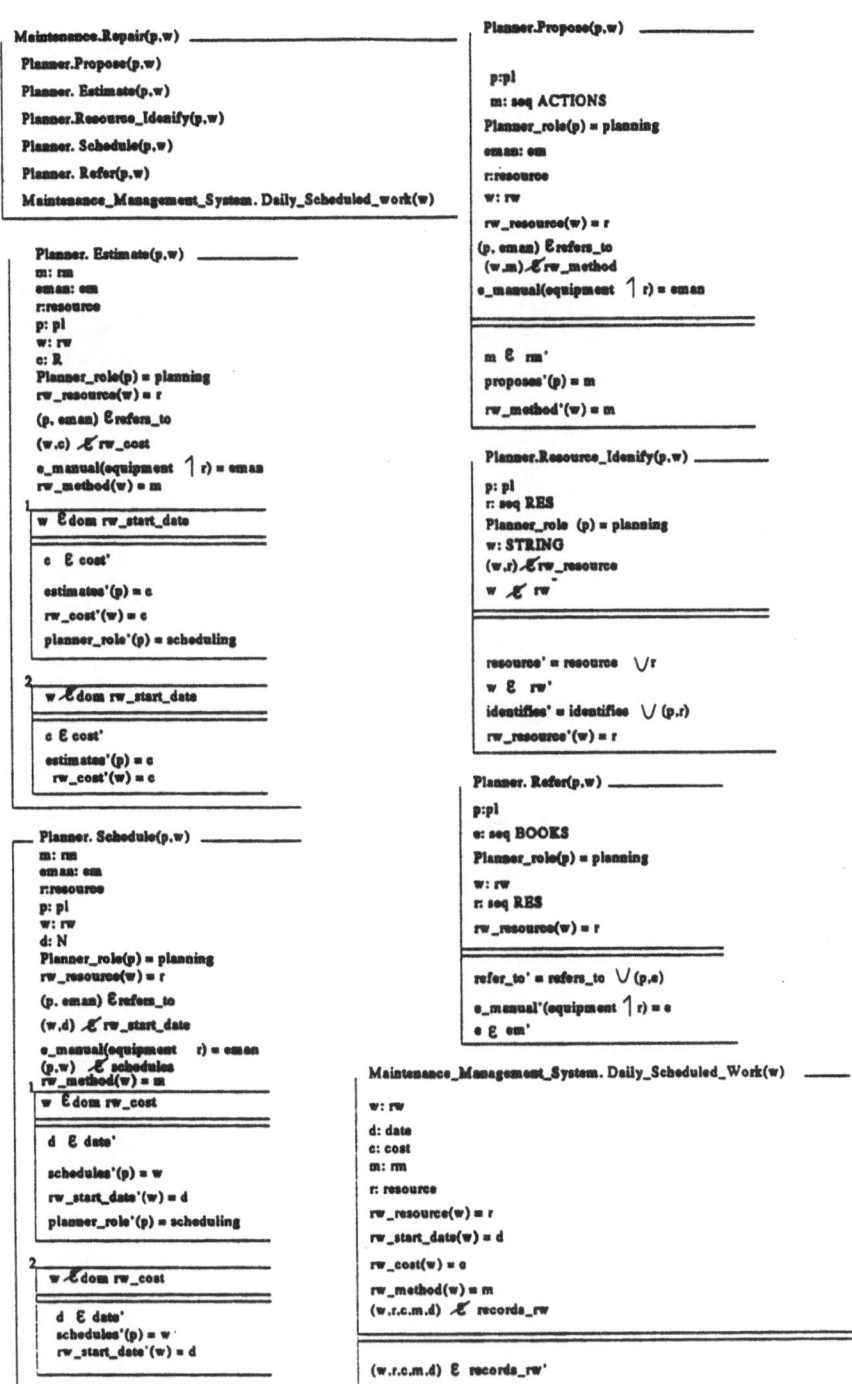

Figure 12. The operation schemata for the planner ALC

101

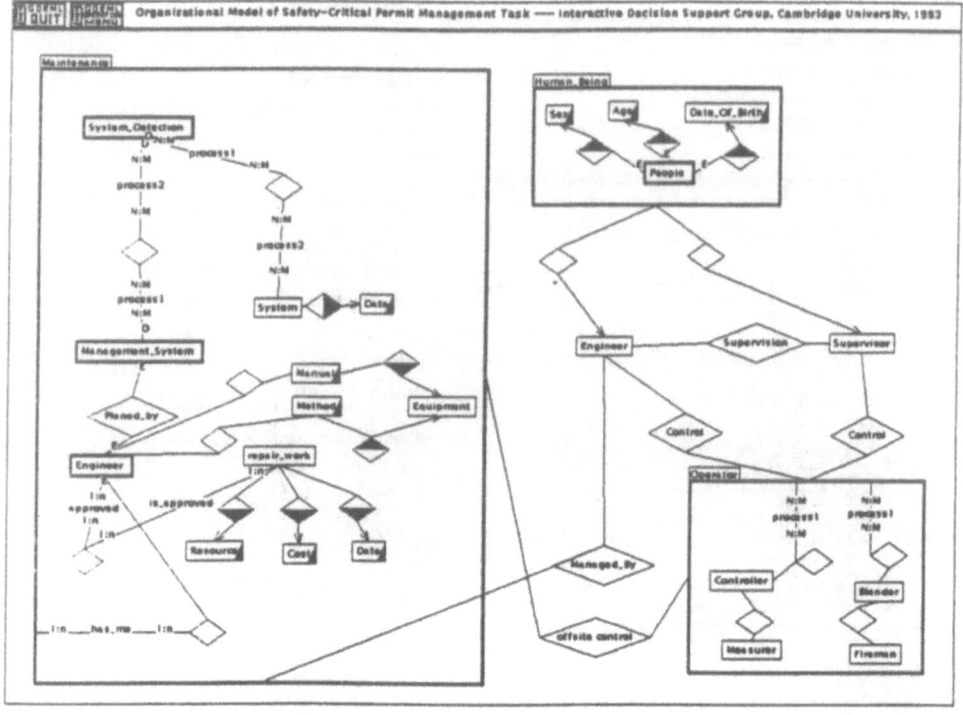

Fig 13: Screen Dump of ORML Editor

The True Cost of Risk and Its Impact on Safety

Dudley Brown and Vivienne Deacon
Alexander & Alexander (UK),

Sophia Langley
University of Birmingham

Abstract

Some form of risk management is in widespread use across industry, but with an emphasis on loss avoidance techniques. Current industrial practice can be characterised and criticised with reference to real life case studies, where the tendency to underestimate the true cost of risk becomes evident. It is desirable to promote risk management as one of a set of safety engineering techniques, and one which is applicable not only as a design aid but also in the post-design stage of a system's life-span to maintain safety standards over time and in the fact of change. To approach risk management from a safety standpoint, methods which yield a more realistic risk costing must be used, and automated decision support systems such as those being designed in the Safety Critical Systems Programme RATIFI Project (Risk Analysis Techniques In Finance and Industry) can contribute towards this.

1. What is Risk Management?

Risk management is an ongoing process which involves the identification and analysis of risk, the formulation of a risk control strategy, and the implementation, audit and adjustment of that strategy over time.

In an informal context, risk management is an activity performed by all good managers when operating in uncertain conditions [1]. To this extent it has not been regarded as a discipline in its own right.

This recognition has been hindered by the fact that the fundamentals of risk management practice have been obscured by the details of the different domains of application, which are enormously diverse, e.g. IT security, air traffic control, financial derivative instruments, toxic waste disposal. Risk management was not recognised as an area of specialist expertise which could be applied across sectors until relatively recently. Witness the fact that the Institute of Risk Management was only established in the UK in 1986.

Technological developments such as advanced control and communications systems now mean that the potential consequences of a single management decision are becoming more and more serious. These new technologies can mean that the effects of a single decision at management level within a large organisation will produce a chain of events around the world within a matter of seconds. The rate of development of these new technologies means that these complex cause-and-effect relationships are often subject to change, so that no one-off risk control strategy will be effective over time [2]

2. Risk Management in Mainstream Industry

It is intended both in this section and in section 4 to relate the experience of the Risk Consultancy Division of Insurance Broker Alexander and Alexander in commercial risk management. Therefore the term "mainstream" is used here to eliminate those industrial sectors where potentially catastrophic safety-critical systems are in use, e.g. those directly involved with military, nuclear power plants, etc. The systems with which Alexander & Alexander comes into contact on a daily basis are ubiquitous and commonplace, but no less hazardous for that.

Most mainstream companies have three types of personnel directly concerned with the management of risk:
1) General management, who are concerned primarily with strategic risk, e.g. will the new changes in product lines lead to a decrease in market share?
2) Financial management, who are concerned with financial risk, e.g. will the rise in value of the pound against the dollar affect this month's export revenue?
3) Technical line-management and safety personnel, who are concerned with technical risk, e.g. will this fault in the manufacturing plant decrease production and cause injury to personnel?

Obviously, these types of concerns overlap, although the extent to which this is recognised varies widely from company to company. (For effective risk analysis, these concerns must be reconciled). Traditionally, a structured approach to risk has only been considered appropriate for technical line management and safety personnel. The main source of general advice external to the organisation has been the insurance sector, i.e. insurers or insurance brokers, whose business involves a unified estimation of risk. Other sources of advice have been regulatory bodies, e.g. the Health and Safety Inspectorate, and those concerned with the legal consequences of risk, e.g. the legal profession and trade unions [3].

Insurance is one of the oldest means of spreading or dissipating risk within a society. An insurer agrees to take on part or whole of a client's risk in return for a premium, usually paid annually. Obviously, an insurer must have sufficient capital to cover the risks he underwrites, and as we have seen recently with Lloyd's of London, disaster results when this equilibrium is disturbed. Wholesale and retail insurance markets have grown up worldwide. The wholesale insurance market is known as the reinsurance market. The tradiational role of the insurance broker has essentially been to find the best insurance deal for his client from the insurers' offers [4].

Accurate risk assessment is the lifeblood of the insurance sector. In practice, this means that it is the insurers and brokers who are the people who have both a vested interest in commercial risk control and the leverage to enforce control measures. This is because industry recognises he need to spread its risks, and insurers can set the price at which they will take on this business according to their assessment of the situation. Insurers and brokers have to go out into industry to check that the risks they have taken on are being controlled by the maintenance of safety standards. They are the people who take on the mundane but essential work such as walking factory floors checking for machinery without guards. They are the people who must be up to date with the latest legislation. They are the people with the financial muscle to oblige management to act on safety. It is common practice for new insurance business to be taken on only upon strict condition that the risks are reduced by the implementation of recommended control measures, which form a customised safety strategy devised for the client by the insurer/broker.

3. Risk Management Is a Safety Engineering Technique

In this section safety engineering is discussed in its broadest sense, so the term 'system' can be defined as equipment, working practices, organisation, etc.,...

Different safety engineering techniques exist to address:

(a) different subsystems of the whole - hardware, software, "liveware", the environmental interface, etc.,...

(b) different stages of the system lifecycle - design, implementation, upgrading, etc.,...

These techniques reduce the risk of malfunction within the subsystem(s) they were designed to address, and at the stage at which they were designed to address it. However, during the system lifecycle, particularly at the on-site operation stage, new risks arise which require a global safety-engineering technique [5].

Risk management is such a technique. Risk management addresses the system as a whole. It is a technique which can be applied at every stage throughout the system life cycle, from design onwards. It is a cyclical process which takes into account changes in the nature of safety threats in the system domain and changes in the structure of the system itself. Moreover, individual techniques do not address risks associated with ineffective management structures, including inadequate communication paths and policy [6].

The role of risk management is based upon the fact that after the application of other specific safety engineering techniques there remains an overall system vulnerability, the "residual vulnerability". This can never be completely eliminated, but it can be minimised. Risk management can be used in the design process to test the residual

105

vulnerability of a design in a "what-if" scenario, but its main effect comes after system installation to keep system vulnerability as low as possible over time and in the face of change [7].

Risk management techniques seek to minimise this residual vulnerability within practical real-world constraints such as financial human and time resources. Rarely are those people investigating and recommending security measures the same people who are empowered to authorise their implementation. Risk management provides a means of communicating and promoting safety standards and recommendations across this technical-commercial divide to those responsible for efficient use of an organisation's resources.

The cost/benefit trade-off is obviously paramount in industrial risk management, where any investment in risk controls must be financially justified. Whilst those in the insurance industry are obviously keen to promote insurance as a risk control measure, it is rarely in the interests of either the broker or the client to use insurance where inappropriate. Brokers are not bound to particular insurers and therefore maintain free status to advise clients as to the available risk control options as independent risk consultants. **It is clear that this role provides one of the most direct platforms for widespread influence of commercial safety standards.**

4. Elements of Risk Management: Risk Identification and Analysis in Industry (see figure 1)

4.1 Risk Identification

The identification of risk is much like a criminal investigation, it requires a probing methodical approach. It is concerned with the perception of risk, i.e. the ability to see a potentially loss-making situation. Risk identification may involve numerous techniques, such as brainstorming, questionnaires, Hazop studies, physical inspections, comparison with historical data, and ranking and rating. It should be noted that the types of techniques used and the extent to which they are applied is governed by the risk management budget.

The manager's role is more predictive than retrospective and future risks, such as impending legislation, or shifting market forces should be considered [8]. There are 3 stages in the process of risk identification:

(i) The identification of a loss-producing event, e.g. personnel dishonesty.
(ii) Analysis of possible operative perils and impinging hazards, e.g.programme tampering.
(iii) The resultant loss effects, e.g. loss of earnings.

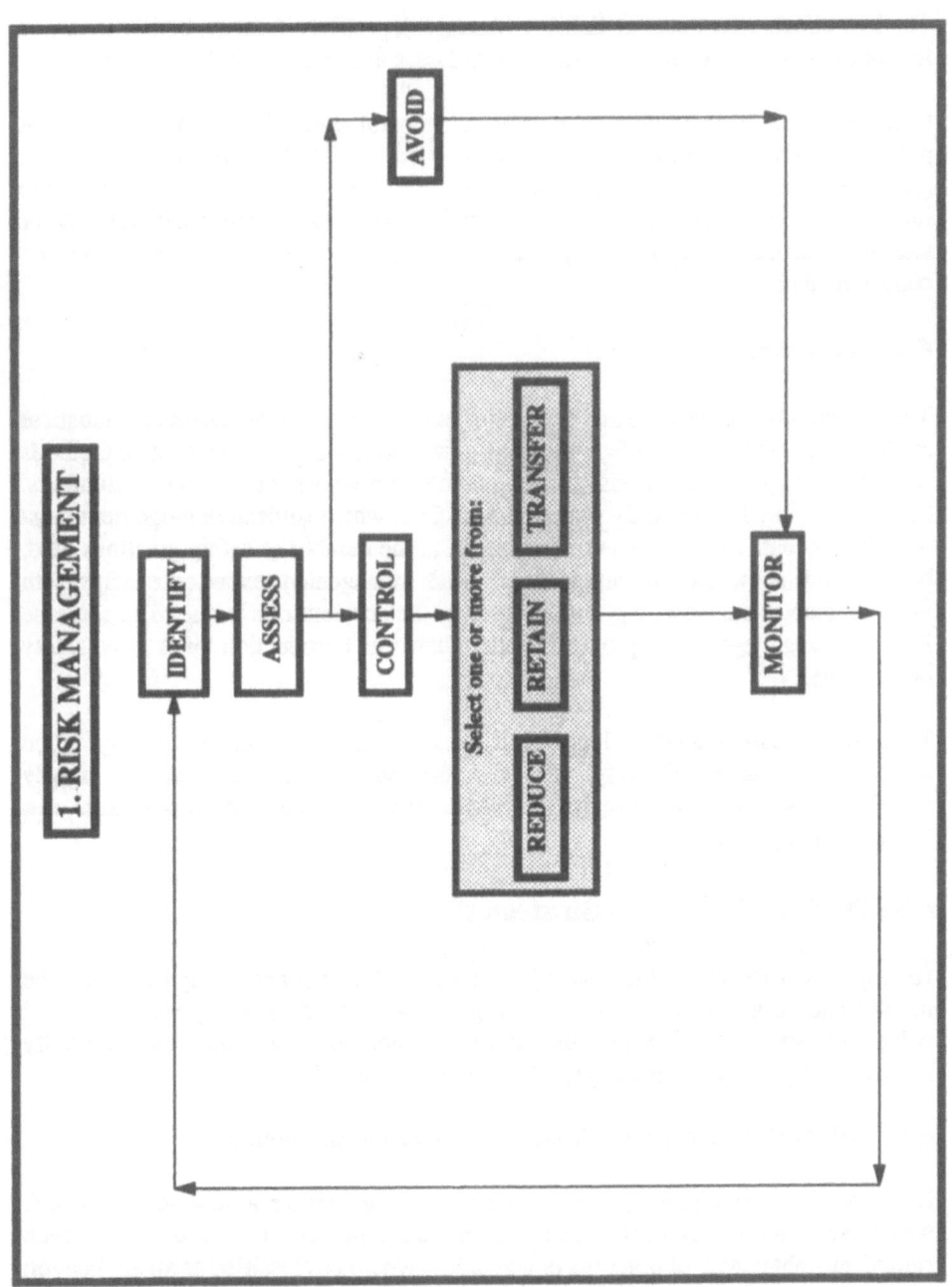

The loss effects may occur as liability (damages), property damage, bodily injury or loss of earnings. Any one event can lead to losses in one or all of these areas.

Collection and interpretation of data is an integral part of the risk management process. Analysis can be undertaken on economic, social or legal grounds. Statistical procedures and probability concepts are used on data collected during the identification stage. It is important for the risk manager to show his financial director that the methods used in his analysis are sound, so as to justify expenditure on loss control methods.

4.2 Assessment

The likelihood of occurrence and severity of outcome can then be assessed. Statistical analysis should be used together with non-quantitative factors. Quantitative methods alone have obvious limitations - few situations can wholly be reduced to numbers. Statistical methods often only narrow the range in which subjective judgement must operate, e.g. data may show an improvement in the number of safety meetings held, but does not show the low attendances. Risk management experience helps with interpretation of statistics, e.g. a sharp drop in the accident rate is due to an increase in the average age of employees, rather than the introduction of a new safety programme.

Traditionally, risk managers have been wary of statistical techniques, relying on subjective estimates. **Clearly, with the increase in complex systems, tightly coupled systems and safety-critical systems, this is increasingly inadequate and therefore dangerous.**

4.3 Which Risk Control Strategies?

Having quantified the risks, expert judgement and cost-benefit analysis should be used to determine which risk control strategies to adopt. There are 4 possible types of risk control strategies. The suitability of one or more to a given situation is partially determined by cost/benefit analysis, discussed in section 5.

4.3.1 Risk Avoidance (Applicable only in prescriptive situations)

Risk Avoidance is a management decision not to undertake a new activity which would create an unacceptable risk to the organisation. Once the activity has been started, any abatement of potential or actual loss would be classified as an application of some other technique. Some examples of avoidance would be:

- rejection of a contract in which the firm assumes responsibility for the risk
- the decision not to manufacture a new product because of danger to the public or high cost or unavailability of product liability insurance.

All other risk control techniques assume that the risk of loss already exists.

4.3.2 Risk Reduction

This considers the reduction of risk within the Company by the development of a programme of loss control. The basic aim of such a programme is to protect the Company assets from wastage caused by accidental loss.

Initially, there is a need to collect data on as many loss-producing incidents as possible, in order to set up an effective programme of remedial action. The first stage of the development will involve the reporting of accidents that result in physical harm to an individual (injury or disease) damage to property, plant, equipment, materials or product, or those "near-miss" accidents where there has been no actual resultant loss, i.e. no injury or damage. .

The second stage of the development towards loss control or risk reduction is achieved by bringing other areas such as fire prevention, security, environmental control, products liability, and business interruption considerations together into one co-ordinated management function with the aim of reducing all accidental losses within the Company's operations.

An example of risk reduction is bunding between oil storage tanks to help to contain fire or spillage. Foam, sprinkler and other fire suppression or detection systems do not prevent fires or explosions, but may reduce the extent and amount of damage which occurs from such events.

4.3.3 Risk Retention

There are two aspects to consider under this heading: risk retention with knowledge and risk retention without knowledge:

(a) With knowledge
A conscious decision is made to retain the risk within the Company's financial operations. This may involve the formation of a captive insurance company or the use of deductibles, i.e. the self-assumption of risk.

Decisions on which risks to self-insure can only be made once all the risks have been identified and effectively evaluated. The Insurers may also force risk rentention on a Company by the imposition of a deductible. This may range from £100 on motor vehicles to £250,000 on the material damage risk in the chemical industry.

(b) Without knowledge
Risk retention without knowledge involves self-assumption via error or omission. This arises because the risks, or their combination and/or aggregation, have not been fully identified and evaluated, and hence the appropriate control action has not been

taken.

Generally, this results in under- or non-insurance. An example of this situation is the Flixborough explosion, where the risk of a vapour cloud explosion was not identified, and hence the subsequent evaluation did not take into account the maximum possible loss from such an explosion. The plant was never fully operational following the loss.

It is generally accepted in risk management that an organisation would retain risks of loss which are not catastrophic, when the cost of losses is exceeded by the cost of transfer. Therefore, sophisticated risk management programs invariably involve some payment of loss which will come out of the current cash flow of the organisation. In order to prevent this retention becoming catastrophic itself, the insurance market provides stop-loss cover both for single loss incidents and in the aggregate.

4.3.4 Risk Transfer

Risk transfer refers to the conventional use of insurance companies in terms of transferring risk from the company to an insurance company by means of risk transfer payment , i.e. the insurance premium.

Other forms of risk transfer also involve forms of contract, but of a non-insurance nature, e.g. the contractual transfer of risk from Lessor to Lessee.

The environment in which an organisation operates is an important consideration in selection of risk management approaches, e.g. a parent company based in the USA insists on certain minimum insurance cover greater than the optimal cover for its UK subsidiary being enforced globally.

4.4 Monitor

Once the risk management strategy has been implemented, it is important to monitor its effect. Accident and incident statistics must be carefully collected for analysis. Note that sufficient time must be given for loss control measures to take effect before evaluating success.

New legislation could make additional risk control measures compulsory, e.g. the recent Management of Health and Safety Regulations call for identification of imminent danger and also health surveillance. The recent hardening of the insurance market has led to large increases in insurance premiums, or reductions in cover, which means that alternative methods, such as increased deductibles (a large excess to be paid by the company) and therefore new loss control methods are introduced.

5. Elements of Risk Management: Cost/Benefit Analysis

Cost/benefit analysis is a decision support tool which attempts to present all the costs and benefits of a particular course of action in a unifying format which allows them to be compared directly.

5.1 Data Required

The Risk Analysis procedure should lead to some means of ordering the relative sizes of risks to the system under study. It may not always be possible to put a quantitative value to some risks. Looking at the components of risk, i.e. likelihood of the event and gravity of its effects, this could be because confidence in the estimates of the likelihoods of the events is too low, or because it is very difficult to quantify the gravity of the event, e.g. how to put a number on the cost of human life. There is obviously often a subjective element to this risk ordering. However, many highly important management decisions (in fact, one could argue, most of these decisions) are made in the light of inexact information. Therefore a subjective management opinion is an element of the risk ordering process this does not invalidate the result, as long as the resulting ordering is subsequently treated as having a subjective element [9].

5.2 Estimating the Costs of Safety

When risks are identified, appropriate safety measures can then be considered. This process is complicated by two factors:

> 1) The mapping of system risks to potential controls is many-to-many
> 2) Potentially hazardous interactions of controls may arise after implementation.

Potential controls should be identified for each risk. There may be several possible different controls for a given risk. A single control may reduce several risks. Each control should be costed. Both direct and indirect costs should be considered in this process. One major issue which may seriously affect the accuracy of the analysis is that of recursion, i.e. the extent to which control measures introduce their own risks. Exact costs may be hard to find, and opinions differ upon exactly what should be costed, but generally this is relatively easy to estimate. (An exception might involve the estimation of the cost of implementing a new working procedure which involved only training and changing attitudes).

5.3 Estimating the Benefits of Safety

This is the hardest part of cost/benefit analysis because estimating the benefits of safety is actually a predictive exercise. The approach most often used is to estimate the reduction in risk or the reduction in loss. In all but the most technically sophisticated environments, the benefits of safety must be expressed in terms of losses avoided. Unless safety benefits are presented in hard financial terms they tend to pale into insignificance in a commercial environment against the hard cost of risk controls.

5.4 The Cost/Benefit Trade-off

This is an optimisation problem where the actual controls to be implemented are selected. Again, quantitative comparisons can form a large part of the decision process, but subjective management opinion will also play a part [10]. Essentially, controls must be chosen so that the benefits of investment in safety outweigh (or, at very least, match) the costs. This can be done with one of several parameters in mind. Many strategies are possible, for example:

> 1) Obtain the highest risk reduction/loss avoidance per pound invested in controls for costs totalling a given safety budget.

This can, of course, lead to a skewed safety profile. It may be that the best value for money can be obtained by eliminating large numbers of small risks, leaving the important ones, which are costly to reduce, uncontrolled.

> 2) Obtain the highest risk reduction/loss avoidance per pound invested incontrols for each risk, starting with the largest and working in descendingorder of importance up to a cost equal to a given safety budget.

This would avoid the problems mentioned in (1).

> 3) Obtain the highest risk reduction/loss avoidance per pound invested in controls for each risk, starting with the largest in the sector of the organisation with the worst risk profile, and working in descending order up to a cost equal to a given safety budget.

> 4) Eliminate certain marked risks which management find unacceptable (these are not just the largest ones - their nature has played a part in the selection, too), then obtain the highest risk reduction/loss avoidance per pound invested in controls up to the costs in the safety budget.

> 5) Reduce risks in all sectors of the organisation to a roughly constant level, which will be determined by the calculation, up to the costs in the safety budget.

> 6) Find the safety budget required to reduce risks in all sectors of the organisation to a level equal, or below, a pre-agreed value of 'acceptable risk'. Rare approach!)

Estimation of risk reduction/loss avoidance is the hardest part of risk management. Even if previous data is available, it is not always reasonable to say that the effect of a particular control measure on risk reduction in one system will be anywhere near the same for the system under study [11].

6. Current Practice Leads to Risk Costing Errors

The identification and quantification of risk is a heavy resource burden for the risk manager. It is time consuming, expensive and there is a dearth of trained experts. Risk is consistently underestimated because only direct costs are taken into account. In the case of large-scale losses, e.g. Piper Alpha, the costs have been well documented, but losses from routine accidents do not tend to be so, e.g. the downtime of a system in one factory leads to lost production, which leads to lost earnings, but in the case where another subsidiary also relies upon that production for the manufacture of its own product, losses snowball in a way which is not documented as well.

In 1989 the HSE's Policy Advisory Unit began a series of five case studies to identify accurately the full cost of accidents. The results led to the conclusion that attention to safety is not only cost effective, but can profit an organisation.

Here is a typical set of indirect costs which often remain hidden in risk identification or underestimated in risk assessment:-

- Safety administration/accident investigation

- Medical centre treatment

- Lost time of injured employees

- Lost time of other employees

- Replacement labour

- Payments to injured employees

- Lost production - business interruption

- Repair to damaged plant

- Replacement of damaged materials

- Insurance excess/deductible

This list is by no means exhaustive but accident costs other than those borne by the company, such as NHS costs, should not be included. Costs should be readily associated with the accident, tangible, realistic and acceptable to management.

6.1 Present Systems of Accident Prevention

Present systems of accident prevention are rendered less effective because:

- line management are not financially accountable for accidents [18].

- very little use is made of economic arguments, relying rather on legal and humanitarian grounds only.

- when an accident occurs in a department, the cost is absorbed into the running costs of the whole site, not the department budget.

- insurance premiums are usually paid by head office, although they may be allocated out to individual locations, but seldom to divisions within a single site.

- in contrast with insurance premiums, safety measures are paid from the department manager's budget, although they are unlikely to be itemised as accident prevention costs.

- most accounting systems mean that it makes no economic sense for the department manager to undertake accident prevention.

- accident costs are often deliberately hidden, to take advantage of maintenance procedures in force in large organisations. Maintenance work tends to be paid by Head Office, rather than the local branch, whereas accident costs come out of the branch budget, e.g. when poorly maintained guttering collapses due to a storm it is really accidental damage, but may be treated as maintenance.

6.2 Co-ordination of External Organisations

The costs and the degree of planning involved in co-ordinating external organisations in the case of a major accident are open to gross underestimation. Those typically involved might be:

114

- HSE Fire Investigator

- Fire Brigade

- Environmental Health

- In-House Auditors

- Police

- Local Authority Building Inspector

- Environmental Inspector - NRA, Local Authority

- Insurers

7. Risk Costing: A Case Study

7.1 Introduction to Risk Costing

The HSE (Health and Safety Executive) case study detailed in this section (taken from The HSE Guidance Note,"The Cost of Accidents at Work", HMSO) exemplifies the problems of risk costing.

Direct costs of risk are relatively easy to assess - the replacement value of equipment, repair costs to premises. The indirect costs of risk, e.g. business interruption, loss of client confidence, litigation, are often overlooked, despite the fact that they are nearly always of a significant size and sometimes are even larger than the direct costs.

Many industrialists and safety professionals have argued that good safety standards can help reduce an organisation's costs when pursued as part of a wider management control strategy. Fisher's work [12], which classified the costs associated with quality and safety functions into prevention, appraisal and failure costs, suggests that even for relatively unsophisticated organisations, failure costs can be greater than the combined prevention and appraisal costs.

Figure 2 shows the costs of a safety programme against the costs of accidents. It suggests that there is a minimum point past which further investment in safety will not be cost-effective. However, few organisations consider that they have reached this point. Control programme costs could include decision making, safety hardware, communication and training time, publicity campaigns, inspection, maintenance, staff. Accident costs could include major and minor personal injury accidents, occupational ill health, equipment and material damage events, product losses, and process and technical breakdowns or damage to the environment [13].

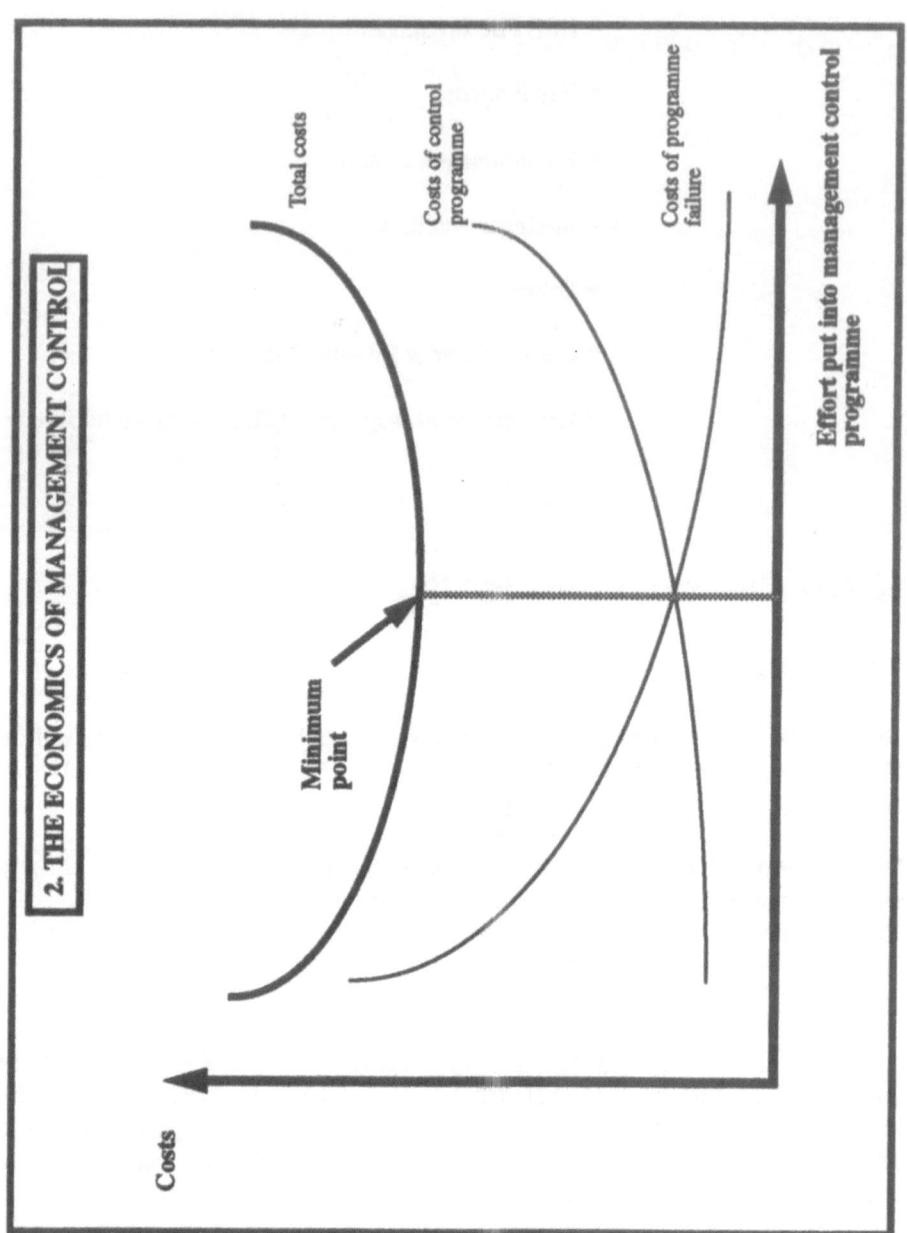

2. THE ECONOMICS OF MANAGEMENT CONTROL

Costs

Total costs

Costs of control programme

Costs of programme failure

Minimum point

Effort put into management control programme

116

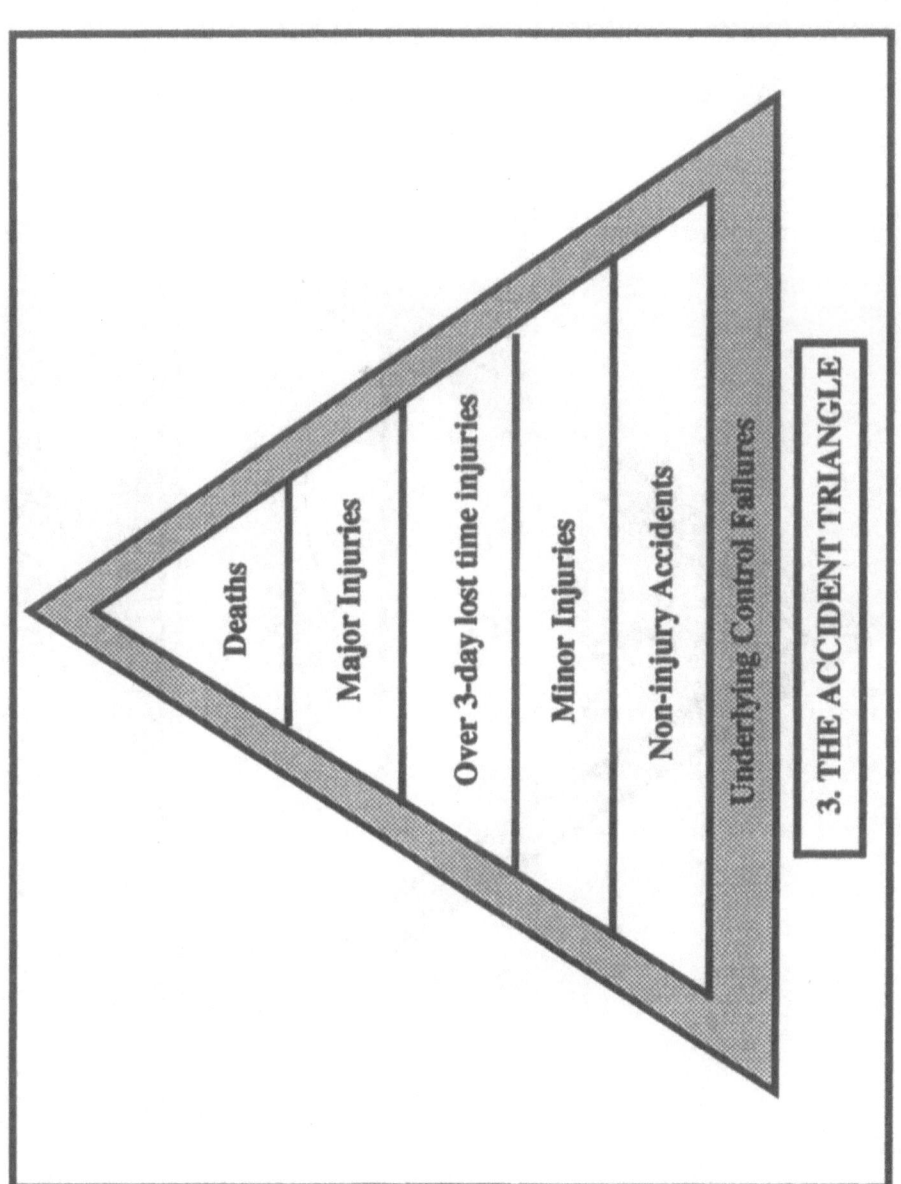

Deaths

Major Injuries

Over 3-day lost time injuries

Minor Injuries

Non-injury Accidents

Underlying Control Failures

3. THE ACCIDENT TRIANGLE

117

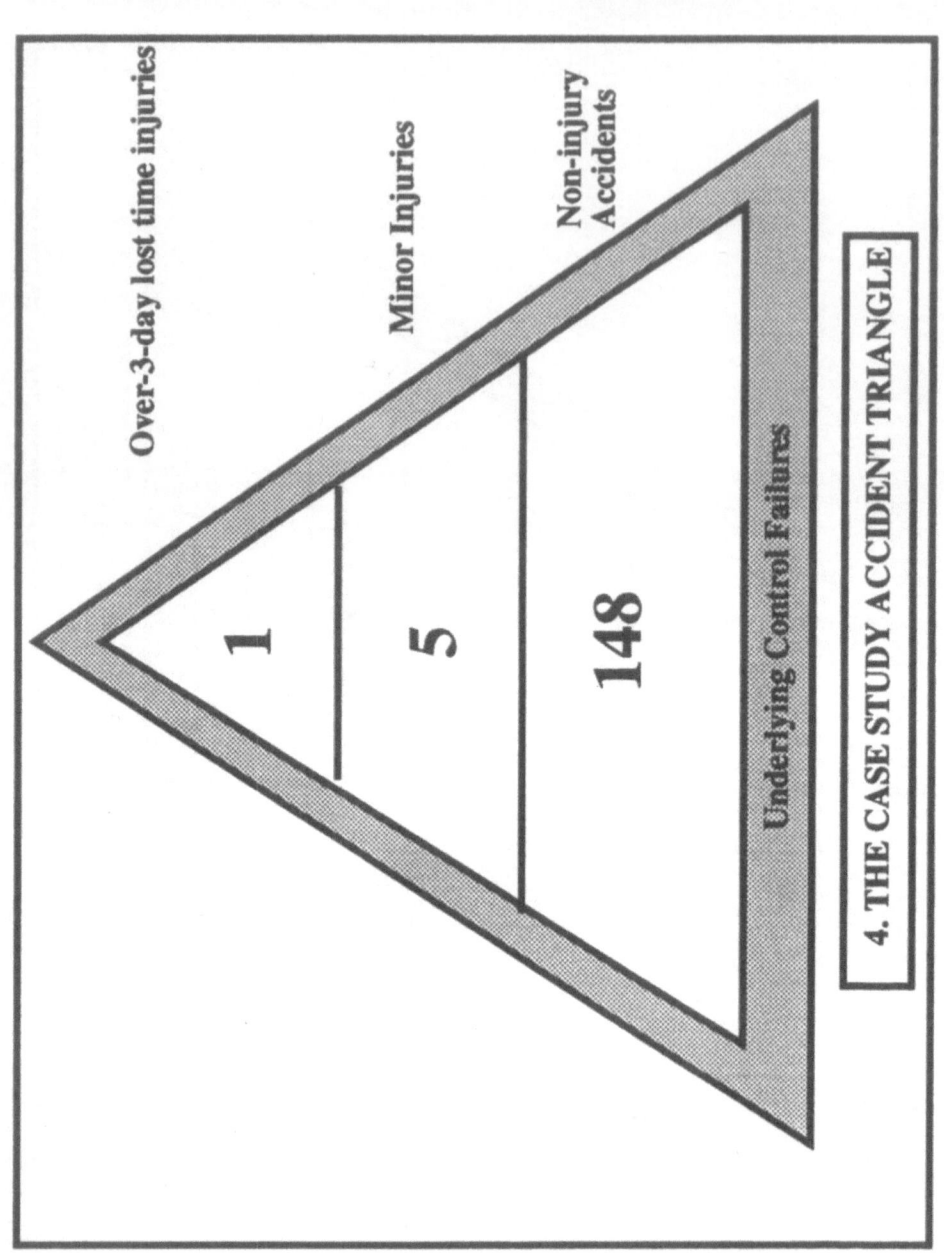

Over-3-day lost time injuries

Minor Injuries

Non-injury Accidents

1

5

148

Underlying Control Failures

4. THE CASE STUDY ACCIDENT TRIANGLE

Total loss control theory uses the triangle concept (see Figure 3) to show that controlling the numerous non-injury accidents will reduce the chances of injuries and fatalities, and so provide proactive control of health and safety [14]. This is based on the belief that, although there is a wide range of immediate causes of accidents in the triangle, the underlying causes are common. The Health and Safety Executive has done research which suggests that there is a broad similarity between the triangles for organisations in the same industry, e.g. construction.

7.2 Case Study: Dairy Product Manufacture

This study took place at one of a number of creameries owned by a leading UK manufacturer of dairy products. 340 people worked at the creamery, organised into product and service departments. Each department had its own supervisors and manager, under the direction of a general manager. There was a range of continuous and batch processes on site and night shifts operated in some departments. During the study the creamery received up to 300,000 gallons on milk per day, collected from over 400 dairy farms, for processing into a range of dairy products. Other work activities included producing colourings, flavourings and packaging for products. Products were sold primarily to large supermarket chains, which imposed strict contractual requirements relating particularly to food quality and hygiene and the timeliness of deliveries. Most of the products had short shelf-lives so there was a continual need to keep machinery and equipment in good working order to maintain production. The site was in the final stages of gaining certification to BS 5750 Part2.

The 13-week study covered periods of high and low demand for the company's products. A 'no-blame' policy was adopted for the duration to encourage staff to report all accidents. A total of 926 accidents were recorded, resulting in financial losses of £184,253. This was equivalent to 1.05% of operating costs during the study period. Opportunity costs, at £59,581, increased this figure to 1.4%. The total costs of all accidents recorded was £243,834. Accidents which cost less than £5 or 15 minutes' lost time were not recorded.

Of the 926 accidents, six resulted in over-3-day injuries and a further 31 injuries and 148 accidents involving damage to property and other non-injury losses were incurred. Ratios between these types of accident are represented in the accident triangle (see Figure 4). The ratio of insured to uninsured costs was 1:36. The difference between this ratio and that recorded in other studies is partly accounted for by variations in insurance carried by different organisations. The important point is that the figure for uninsured costs far outweighed that for insured costs.

Over £3000 worth of losses resulted from 3 separate occasions of bacterial contamination of equipment used for handling the company's product. All the contaminations were identified and contained, but these accidents had the potential for major losses if public health had been affected.

£2,000 worth of damage was caused when a seal of the correct size but wrong specification was fitted to a machine following a breakdown. In other circumstances of non-adherence to specification, major accidents have resulted.

£8,200 worth of losses was caused when blocked bag filters became over-pressurised. Five tonnes of milk powder were lost in the exhaust air. Whilst a dust explosion did not occur and no one was injured, the risk of a milk dust explosion had not been fully assessed.

£1800 worth of damage was caused when a road tanker was driven away while still connected to the factory pipework. Management had not identified the full potential of this hazard nor assessed the risk involved. If, for example, a flammable liquid had been involved rather than the company's comparatively innocuous product, there could have been a major accident involving the risk of fire and explosion.

A large number of accidents happened around the start of a shift at 6 am. However, the electricians and mechanics needed to carry out repairs were unavailable until 8 am because of the nature of the company's shift system. Following presentation of the study results to the company, this problem was largely resolved, by providing key machine operators with further training so that they could carry out some of the routine preventative maintenance tasks themselves. This subsequently considerably reduced the number of accidents and lost time experienced by the company following the study.

The lessons from this particular study could be applied to some chemical companies, as many of the processes undertaken at the factory were similar to those carried out in chemical production plants, although the nature of the materials handled would be different. The potential for major catastrophes did exist, however, e.g. the risk of fire or .explosion during the handling of powdered milk and the possibility of microbiological contamination of both ingredients and products with the potential for a food poisoning epidemic.

Identified costs included:

- Employers, product and public liability

- Property damage

- Material damage

- Business interruption

- Product and material damage

- Plant and building damage

- Equipment damage

Unidentified costs included:

- Legal costs

- Expenditure on emergency supplies

- Clearing the site
- Production delays

- Overtime working and temporary labour

- Investigation time

- Supervisors' time (diverted)

- Clerical effort

- Fines

- Loss of expertise/experience

The fact that there were £30 worth of unidentified costs for every £1 pound of identified costs in this study serves to illustrate how serious the gravity of the hidden costs can be in practical situations (see Figure 5).

Organisational structures themselves can lead to inaccurate risk costing, an issue touched upon in section 6 on present systems of accident prevention. This is because of accounting practices common within large organisations. Maintenance costs will usually be paid by Head Office, whereas accident costs and safety controls will be assigned to local budgets. This discourages accident reporting - if accident costs can be viewed as maintenance requirements they will not have to be paid for locally. There is therefore no motivation for investment in safety controls at local level. **This accounting structure leads to widespread lethargy on safety and is responsible for lowering safety standards.**

8. Good Safety Policy is Built upon True Risk Costings

Existing legislation already promotes a structured approach to risk management. For example, the 1974 Health and Safety at Work Act stipulates that employers must compute risks against the cost and effort involved in control measures. COSHH regulations require identification of risk to determine control measures. The 1972 Robens report on health and safety suggested better knowledge of accident costs would lead to more informed decision making and that accident costs should be displayed on the balance sheet, so the same effort towards reduction could be applied

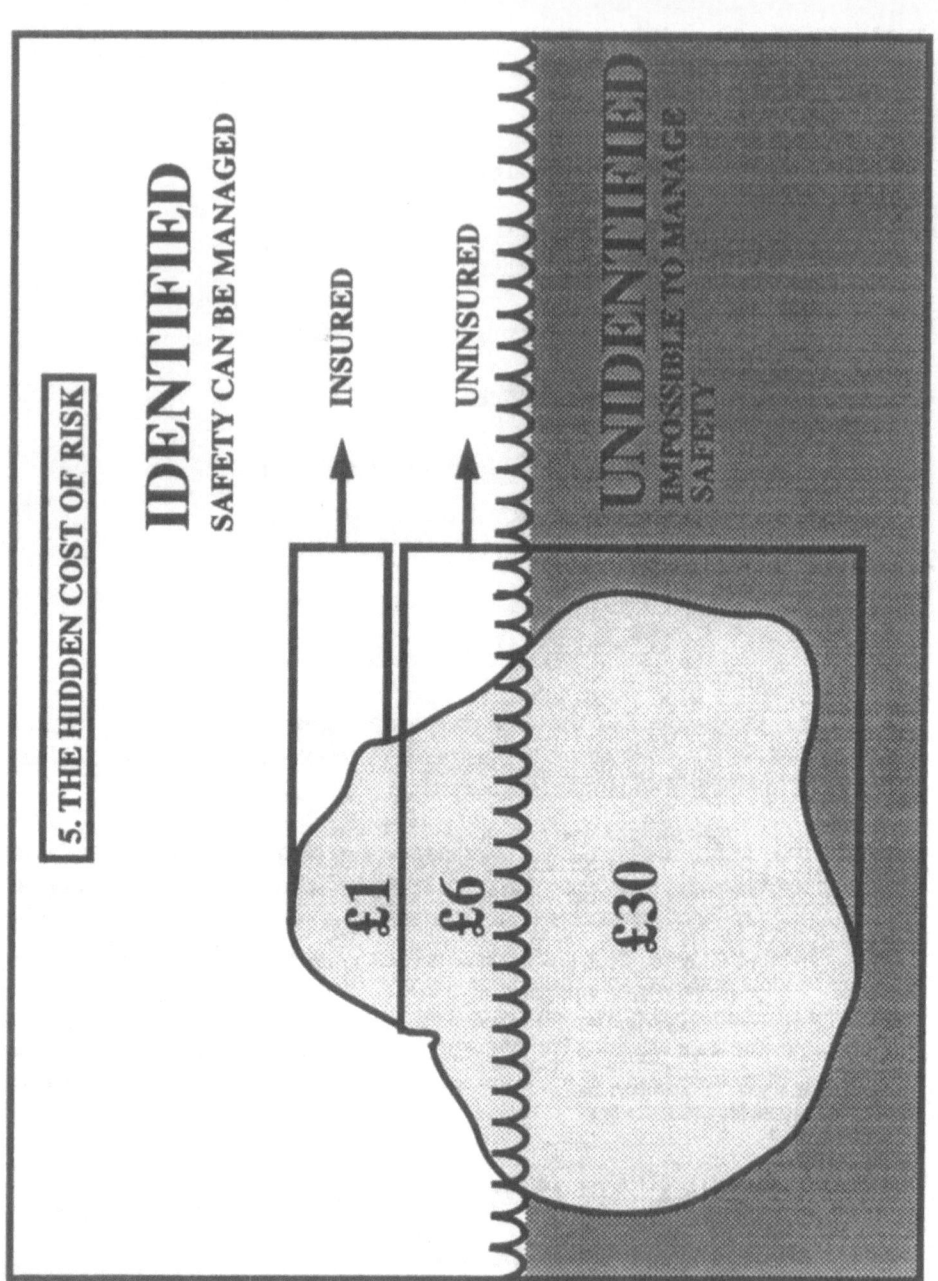

5. THE HIDDEN COST OF RISK

IDENTIFIED
SAFETY CAN BE MANAGED

INSURED

UNINSURED

UNIDENTIFIED
IMPOSSIBLE TO MANAGE
SAFETY

£1

£6

£30

as towards other facets of the business.Cost/benefit analysis is now being used by the Health and Safety Executive (HSE) to justify new legislation.

Within the structures of our risk management model in section 2 the obvious ways in which miscalculation can occur are:

i) Identification - risks may not e fully identified due to cost, time constraints, lack of understanding, ...

ii) Assessment - the effect of risk may not be fully understood as indirect and/or hidden costs have not been included in the calculation, i.e. the cost of accidents is underestimated.

iii) Control - during cost/benefit analysis, after control measures have been costed, erroneous decisions not to introduce given controls are made because investment in control seems non-viable when compared with the cost of accidents. Yet the cost of accidents may have been underestimated as in (ii). Therefore control measures which are in fact cost effective are not introduced.

iv) Control - the initial choice of risk control strategies can be made due to incorrect and/or insufficient data.

It is apparent that more accurate risk costing techniques must be applied in order to justify the resource investment required to implement the recommended structured approach to risk management. Only then will this approach yield the benefits of improved safety standards.

9. Poor Risk Costing Prejudices the Safety Case

Poor risk costing is having a serious effect on safety standards by prejudicing the Safety Case. When the cost of an accident is underestimated, then the risk of the accident is underestimated. This is because risk is calculated in terms of the gravity and the likelihood of an event, and gravity is frequently considered as cost alone.

In the cost/benefit analysis of safety controls, the benefit of safety is the reduction of risk [15]. If risk has been underestimated, then the benefits of safety controls will be underestimated in turn. The cost of safety is the cost of risk control measures. For a given accident, the cost of appropriate safety measures will appear disproportionately high relative to the benefits when costings have been inaccurately low. The net result is that managers will be biased against implementing appropriate safety measures if they are using cost/benefit analysis on safety (see Figure 6).

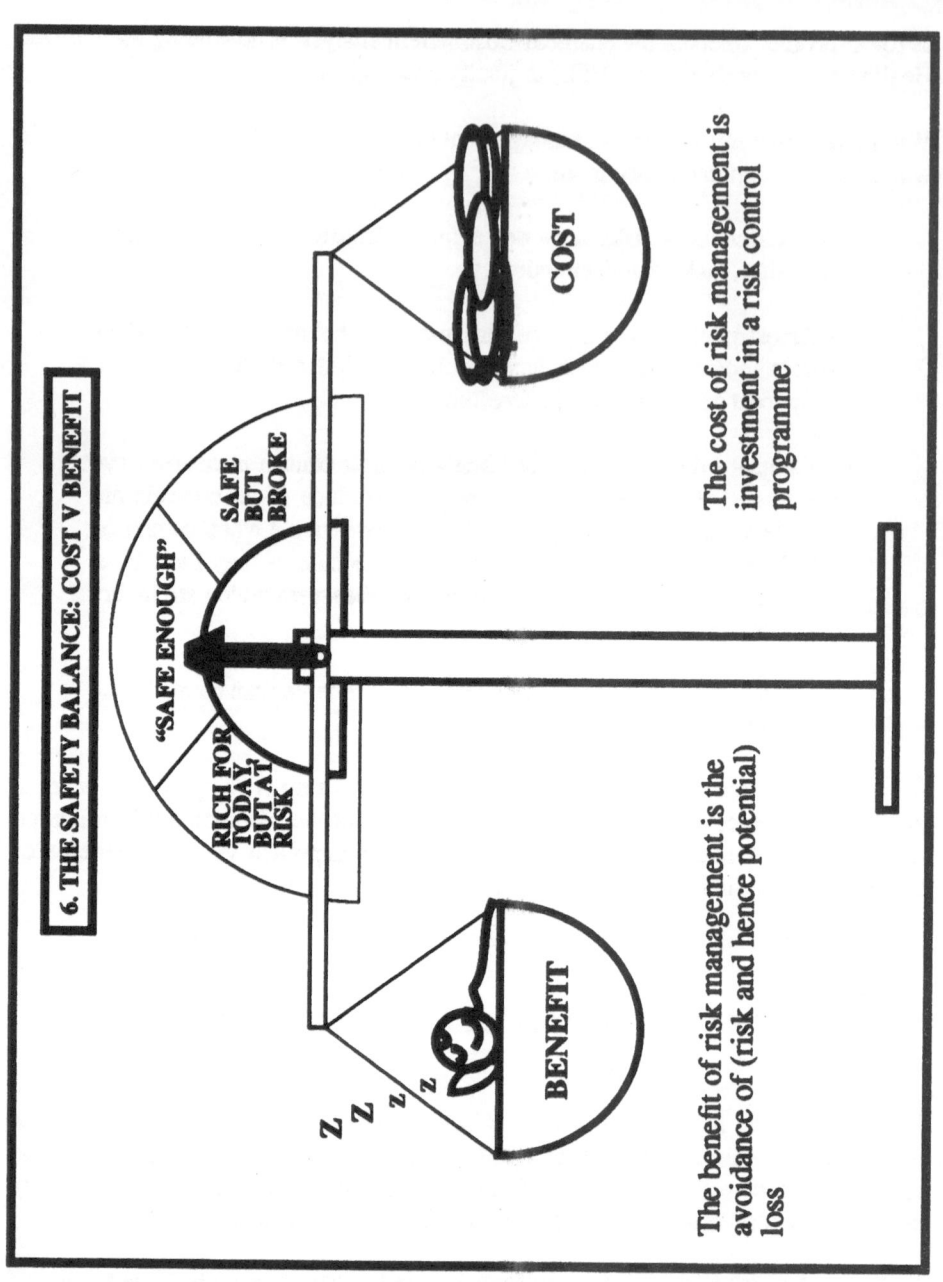

6. THE SAFETY BALANCE: COST V BENEFIT

"SAFE ENOUGH"

SAFE BUT BROKE

RICH FOR TODAY BUT AT RISK

COST

BENEFIT

The cost of risk management is investment in a risk control programme

The benefit of risk management is the avoidance of (risk and hence potential) loss

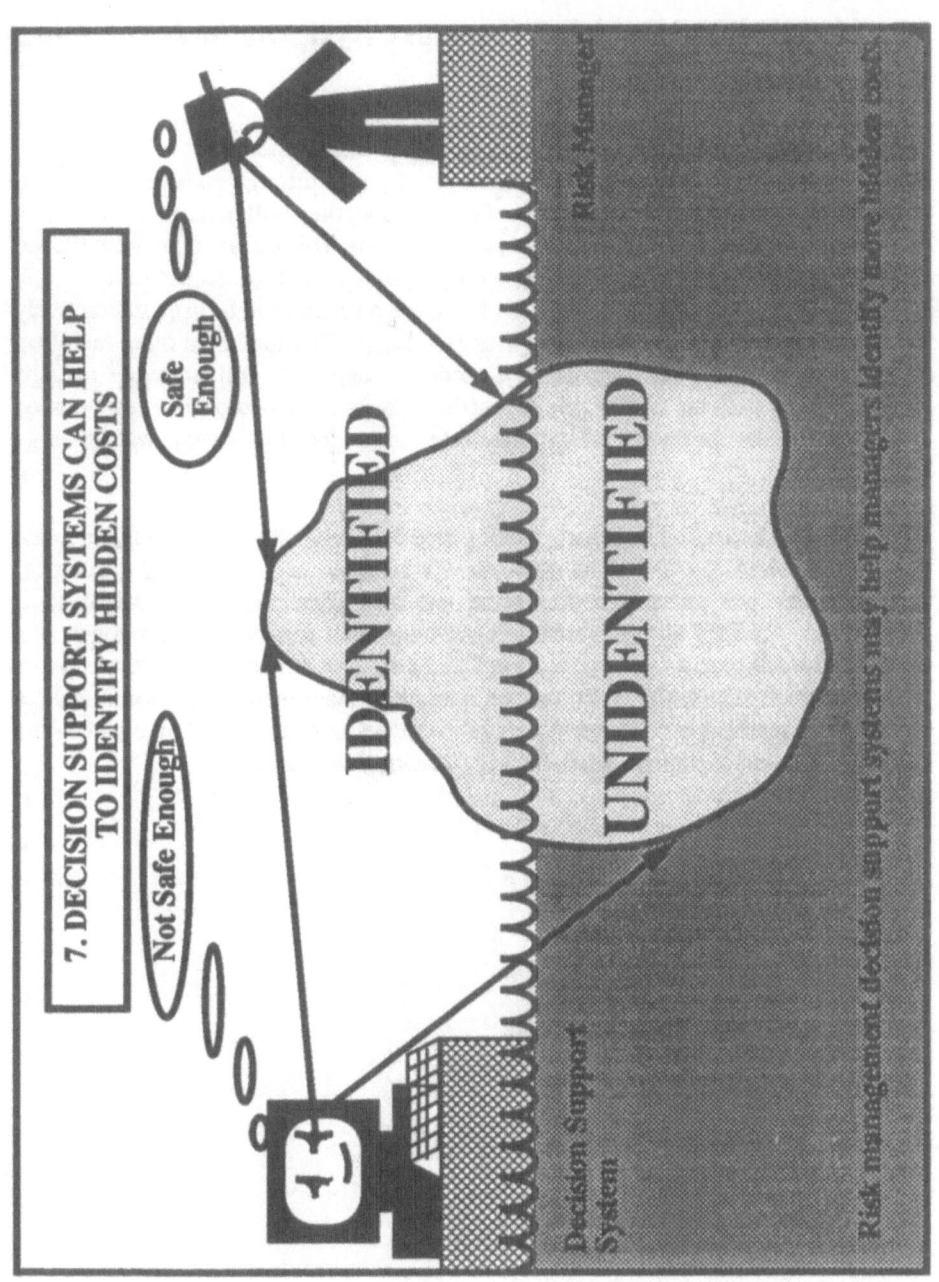

7. DECISION SUPPORT SYSTEMS CAN HELP TO IDENTIFY HIDDEN COSTS

IDENTIFIED

UNIDENTIFIED

Safe Enough

Not Safe Enough

Risk Management

Decision Support System

Risk management decision support systems may help managers identify more hidden costs.

10. The Role of Decision Support Systems in Improving Risk Costing

Managers must be presented with accurate information upon which to base their decisions on safety. If managers are taken step by step through consideration of all potential elements or types of elements of risk costing, they will obtain a standard and more accurate assessment of the true cost of risk [16]. This can be done with the aid of decision support systems which can store knowledge on hazards and controls and on the elements of cost/benefit analysis [17]. Such a system can be used interactively by a manager to guide his safety decisions (see Figure 7). Input from other members of the organisation with financial and technical expertise will probably also be required. Such systems will enable managers to make better-informed decisions on safety, and thus promote safety standards from the top down within large organisations.

Alexander & Alexander UK is participating with the University of Birmingham in the newly-established RATIFI (Risk Analysis Techniques in Finance and Insurance) project, which was granted funding under the DTI/SERC Safety-Critical Systems Programme. RATIFI aims to promote a structured risk management approach as a safety engineering technique, and to design and prototype suitable support tools for its implementation in industry. A successful attempt to meet these goals would make a significant contribution to the problems of risk costing discussed above, and to the quality of automated support available to practising risk managers.

References

1. Carter RL. Handbook of risk management. Kluwer-Harrapp, 1992
2. Perrow C. Normal accidents, Living with high-risk technologies. BasicBooks,New York, 1984
3. Gordon A. Risk and the business environment. Witherby & Co, London, 1987
4. Crockford N, Administration of Insurance. Witherby & Co, 1987
5. Kuhlmann A. Introduction to safety science. Springer-Verlag, New York, 1986
6. Horton FW, Lewis D. Great information disasters. Aslib, London, 1991
7. Cox SJ, Tait NRS. Reliabilty, safety and risk management, An integrated approach. Butterworth Heinemann, Oxford, 1991
8. Casti J. Searching for certainty, What science can know about the future. Scribners, London, 1992
9. Granger Morgan M, Henrion M, Uncertainty, A guide to dealing with uncertainty in qualitative risk and policy analysis, Cambridge University Press, Cambridge, 1990
10. Hastings WJ. Business finance for risk management. Witherby and Co, London, 1987
11. Bannister JE, Bawcutt PA. Practical Risk Management, Witherby & Co, 1981
12. Fisher t, A. 'quality' approach to occupational health, safety and rehabilitation. Journal of Occupational Health and Safety 1991, 7:23-28
13. Bird FE, Loftus RG. Loss control management. Loganville, Georgia Institute Publishing, 1976
14. Bird FE, Germain GL. Practical loss control leadership. Loganville, Georgia Institute Publishing, 1985
15. Royal Society Study Group. Risk assessment. Royal Society, 1983
16. Fox MS. Enterprise modelling: issues and directions. In: Avignon '93. Proceedings of the Thirteenth International Conference on Expert Systems and their Applications, Avignon, France, EC2, 1993.
17. Spiegelhalter: Uncertainty management in expert systems. In: Modelling Uncertainty. Cambridge Programme for Industry, 1992
18. Bamber L, Accident Costing, Alexander and Alexander (UK), Ltd.

Risk and Safety Reviews

Geoff Wells and Mike Wardman

(Dept of Mech. and Process Eng, Sheffield University)

1. Synopsis

A considerable number of safety reviews are carried out during the life of a plant. Some of these, particularly HAZOP, have become key milestones during project development and are virtually mandatory in order to demonstrate successful safety management of a project and plant. This study relates the timing of safety reviews to the stages in a project, hazard analysis methods, risk assessment methods, process safety sociotechnical system reviews and safety audits. An indication is given of current values of tolerable risk and a typical short cut risk assessment following Preliminary Hazard Analysis is illustrated. The effect of management and other root cause factors is demonstrated by the adjustment of average case values. The study shows how increasingly such methods of risk assessment can be used to generate helpful information to assist in risk management throughout the life of the plant. The importance of maintaining a correct model of the incident scenario is emphasised as well as evaluating other uncertainties in risk estimation.

2. Hazard Analysis Methods

Process safety reviews interact with a project from its inception through to its commissioning, see Figure 1[1].

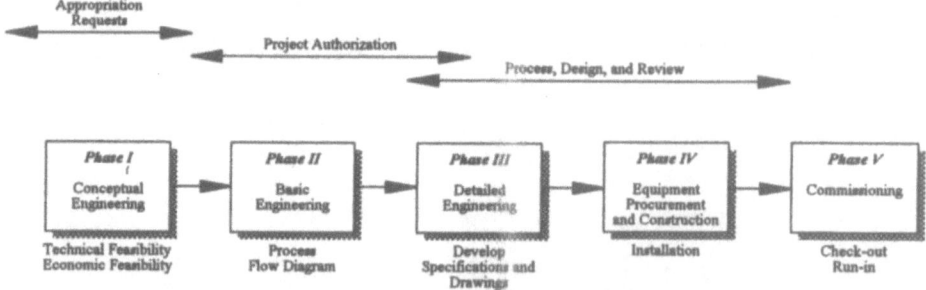

Figure 1: The Phases of a Capital Project

There is an overlap between phases and on certain projects the phases are combined. Thus if the process technology is well-known then Phase I and Phase II are combined. Analysis also continues throughout the life of the plant with modifications being treated as individual projects.

A *Hazard Analysis* involves the identification of undesired or adverse events that lead to the materialisation of a hazard, the analysis of the mechanisms by which these undesired events could occur and usually the estimation of the extent, magnitude and likelihood of any harmful effects. It is in theory applied only to the identification of hazards and the consequences of the credible accident sequences of each hazard. Unfortunately non-credible accidents have a habit of occurring. Major incidents usually occur when a significant release of material has an impact on people, the environment or the business. A model of the development of an incident from its root causes is indicated in Table 1. This is invaluable in carrying out Hazard Analysis and Risk Assessment.

Hazard Analysis is dominated by safety reviews. These normally signify the progress of a capital project on its way to completion. The hazard reviews which are commonly carried out by industry are named variously in different countries and also vary according to need and information available. Consequently here there has been an attempt to standardise the terminology.

Figure 2 illustrates the link between safety reviews and the incident scenario. Descriptions of the main safety reviews now follow.

Concept Hazard Analysis, CHA, is based on the use of keywords such as toxics, flammables, temperature, pressure, drop/fall etc. to identify hazards and general deviations from good practice. The basic *Concept Hazard Analysis Initial Review, CHAIR,* is carried out as specified by the project review. It corresponds to HAZOP I in the terminology promoted by Turney[3] amongst others. What is here termed *Concept Hazard Analysis, CHA,* has been developed as an extension of this procedure[4] and involves some identification of protective measures, consequences and generation of safeguards.

Preliminary Hazard Analysis, PHA, can be used to prioritise risk studies in order to be reasonably assured that no major problem will arise in subsequent detailed engineering. Conventionally this has focused around the release of material but a much better starting point is at a dangerous disturbance of plant, such as a dangerous temperatures and pressures, deterioration of materials of construction, an available opening to atmosphere, change in a planned discharge etc. The aim is to identify all major events but not all their immediate causes. The method requires a better determination of the incident scenario than a CHA and hence is carried out at a later stage of the basic engineering.

Preliminary Consequence Analysis, PCA, is used to assess the consequences of a release in terms of severity categories. It is only carried out for some releases as others can normally be estimated from the results. If members of the public or escalation by domino effects are involved it is necessary to carry out a full consequence analysis to ascertain societal risk and additional information is received before the evaluation of harm and damage can be estimated.

Hazard and Operability Studies, HAZOP, is a technique to identify process hazards and potential operating problems using a series of guide words to study hazardous deviations. The aim is to generate most immediate causes of deviations and to identify

Table 1. General Incident Scenario, Wells[2]

IMPACT: HARM AND DAMAGE
- Consequences categories (Appreciable to Catastrophic), see Allum[5].
- Harm to environment
- Harm to public or workforce
- Damage to property or plant
- Loss of use or occupancy
- Damage to business and business interuption
- Minor consequences / Near-miss

FAILURE TO PREVENT FURTHER ESCALATION
- Post incident emergency response inadequate
- Emergency response inadequate
- Failure to prevent further escalation by fire, explosion or toxic release
- Secondary loss or spread of process material

FAILURE TO MITIGATE OR AVOID ESCALATION
- Failure to avoid a toxic release
- Escalation by fire or explosion
- Ignition of flammable mixture
- Release fails to disperse or attenuate rapidly or it accumulates
- Response by countermeasures inadequate or failed
- Immediate response to release inadequate

SIGNIFICANT RELEASE OF MATERIAL
- Release capable of significant harm
- Failure to recover the situation
- Release of material by rupture or discharge

DANGEROUS DISTURBANCE OF PLANT
- Rupture on exceeding mechanical design limitations
- Rupture within normal operating conditions
- Flow through abnormal opening to atmosphere
- Adverse change in a planned product or other release
- Entry into vessel

FAILURE TO CONTROL THE SITUATION
- Emergency control systems fail to correct the situation
- Normal control systems fail to correct the situation
- Operators fail to correct the situation
- Maintenance fail to correct the situation

HAZARDOUS DISTURBANCE OF PLANT
- Deviations in pressure, temperature, overload, stress
- Deviations in flow of material and other parameters
- Construction defective or deteriorated in service
- Abnormal opening in equipment
- Adverse change in planned product or other release
- Change in mode of operation

IMMEDIATE CAUSES OF INCIDENT
- Deviation at the human-machine interface
- Defects directly cause loss of plant integrity
- Technical deviation in equipment, feeds or services
- Technical deviation in the control system
- Change from design intent
- Environmental and external causes of disturbance

ROOT CAUSES OF INADEQUATE SOCIOTECHNICAL SYSTEM
- Inadequate engineering integrity
- Inadequate operator performance
- Inadequate working environment
- Inadequate procedures and practices
- Inadequate communication and information
- Inadequate management control
- Inadequate site and plant facilities
- Inadequate organisation and management
- Inadequate system climate
- Inadequate external systems

Figure 2: Safety Reviews and the Incident Scenario

Incident Scenario		Concept Studies	Preliminary Studies	Detailed Studies	Special Studies
Risk Evaluation		Hazard Indices	SCRAM	QRA	
Major Events	Mitigation Protection				Emergency Plans
Release of Material		CHAIR	Preliminary Consequence Analysis	Consequence Analysis	Critical Examination
		Concept Hazard Analysis		Safety Report	COSHH
Dangerous Disturbance	Recovery Action		Preliminary Hazard Analysis		HTA/HAEA
Emergency Control				Operability review	
				HAZOP	FMEA/HTA
Hazardous Disturbance	Normal Control			Process Audits	HAEA
Immediate Causes				Safety	
Root Causes		Project Review	PSSSR	PSMSA	

CHAIR   Concept Hazard Analysis Initial Review
PSSSR   Process Safety Sociotechnical System Review
PSMSA   Process Safety Management System Audit
FMEA   Failure Mode and Effect Analysis
COSHH   Control of Substances Hazardous to Health

HTA   Hierarchical Task Analysis
HAEA   Human Action Error Analysis
SCRAM   Short-Cut Risk Assessment Method
QRA   Quantified Risk Assessment

131

consequences. There is considerable emphasis on the effectiveness of emergency control systems but little on harm and damage. The method was designed to be carried out at a specific time when the detailed P&I diagrams were available for checking. Detailed design changes are not normally expected at this stage.

Failure Mode and Effect Analysis, FMEA, is a hazard identification technique in which all known failure modes of components or features of a system are considered in turn and undesired outcomes are noted. The system has had limited use in the chemical industry in Europe as it is tedious and does not readily identify composition changes. Data for reliability studies can be very difficult to obtain.

Task Analysis is a systematic method for analysing a task into its goals and the actions and plans required to achieve these goals. A system is generally broken down into the various steps of the operation or process. The method is often extended to evaluate human reliability by various methods and in particular by Human Action Error Analysis, HAEA, which enables the analyst to identify the external error modes that may occur at a task step and identify both their root causes and recovery mechanisms. A range of hybrid schemes give a better integrarion with HAZOP studies.

What-if Analysis asks various questions prefaced by the phrase "What-if ". As such it can be used at any stage during the life of the plant and for audit studies. It can be used as a general check on other safety reviews, as the question draws attention to whether a particular facet has been studied. Also each question usually generates further queries when the study is used during a meeting format. The procedure is currently being used more frequently than hitherto in the UK.

An *inherently safe and user-friendly plant* is a desirable objective. At some stage during the project it is necessary to have a brainstorming session to generate modifications to the process method. This can be structured in the form of a *Critical Examination*[4] of the system using What-How-When-Where-Who-Why questions. The emphasis must be on what can we do better as well as what can be done worse. Not only does this lead to better safety but can result in considerable improvements in performance. It can be undertaken at around the Concept Hazard Analysis but is probably of greater value during Preliminary Hazard Analysis. It is regularly carried out during Task Analysis.

3. Process Safety Sociotechnical System Studies

The sociotechnical approach emphasises the individual social, organisational and managerial aspects which influence human behaviour and ultimately affects the systems' performance and its safety. In our work we tend to segregate the design issues out seperately exceppt when carrying ou t incident studies.

A *safety audit* is a thorough examination of all, or part, of a total operating system with relevance to safety. Many specialist systems exist and it is of benefit to tailor a particular method to the needs of the user.

Process Safety Management System Audits, PSMSA, can be defined[2] as a review of the mechanisms a company has developed to provide increased assurance that its operating units have appropriate systems in place to manage process risks. Such an audit examines the process safety make-up of operations including: policies and procedures, management organisation, the planning process, risk assessment and risk management activities, management information systems and internal reviews/inspection programs. Some version of audits allow a ranking to be assessed which enable the values calculated for residual risk of a plant to be adjusted.

Process Safety Audits, PSAudits, are intended to provide management with increased assurance that operating facilities and process units have been designed, constructed, operated and maintained such that the safety and health of people and environment are properly protected. Such audits should also include an element for the upgrading of previous risk evaluations as external factors and operating data change and procedures for the evaluation of risk improve.

Environmental Audits are being carried out which not only consider the effect of the release of material on the natural environment but examine the impact of the whole plant on its environment and identify whether the products and route are in conflict with the 'green' objectives and policy of the company. We believe safety, health and environment should not be separated out when considering hazards.

Compliance reviews are used to confirm that a facility's operations comply with applicable laws and regulations. Such reviews can be designed to indicate compliance with industry or associated standards.

A *Process Safety Sociotechnical System Review, PSSSR,*[5] can be used to evaluate the impact of a new capital project on an existing system. Keywords are used to generate discussion in key areas. This has been found to be not completely satisfactory as staff are not as familiar with factors resulting in plant problems. Consequently for a Process Safety Sociotechnical System Analysis, PSSSA, the keywords are augmented by a list of preconditions for failure. The study is best applied to specific incident scenarios.

An *Operability Review* can be used to review how the operators and maintenance staff will run and service the plant. The assumptions included in the hazard analysis and risk assessments should be practicable. The procedures and operating instructions should be examined, incorporated in training and then reviewed prior to commissioning. It is essential to check the recovery or emergency preparedness at each adverse event. It can be readily incorporated within the training programme.

Precommissioning Studies (also known as Pre-startup Studies) are forms of process safety audits. The precommissioning review is carried out prior to the introduction of materials, and is used to check that actions from hazard studies have been incorporated and installed and all instructions and procedures comply with any requirements identified by previous hazard stages and are satisfactory for the safe operation of the plant. A precommissioning inspection checks compliance with company and legislative requirements.

An *Operating Review*, known as HAZOP VI in ICI[3], is carried out after the establishment of beneficial production in order to identify and record operating and maintenance difficulties and ensure feedback to engineering. Subsequent reviews can be undertaken as part of the Safety Improvement Programme adopted by the company to cover all or part of a facility.

A *Safety Report* is the presentation of a justification for the safety of an installation. In the UK this is used in connection with the CIMAH Regulations. The report aims to identify the nature and scale of the use of a dangerous substance at an activity; to identify the type, relative likelihood and consequences of major accidents that can occur; to give an account of the arrangements for safe operation of the activity for the control of serious deviations that could lead to a major accident; and to give an account of emergency procedures at the site. The manufacturer must satisfy the Regulator that all major accident hazards are identified and the proposed or actual precautions are as described and appropriate to the hazards.

A *Safety Case* is more extensive than a Safety Report and comprises the documents available within the Company that show whether overall a sufficient case for safety has been made. At present this case is examined by the Offshore Safety Division of the HSE and their acceptance is the beginning of a process, not the end: the documents become essential source documents for OSD inspectors.

Incident reports should be made on all near-miss events and accidents. The report should not be geared solely to identifying the immediate cause of the accident and assigning blame. The aim is to generate a complete picture of the incident scenario with the identification of root causes including managerial and organisational failings which, by definition, are certain to arise except for some very exceptional external causes.

Many other studies concerning specific safety features continue throughout the life of the plant. Particular note should be made of studies associated with Task Analysis and the Control Of Exposure to Substances Hazardous to Health, COSHH.

4. The Timing of Safety Reviews

The timing of hazard reviews is summarised in Table 2. This diagram should be treated as approximate only. Such timing varies considerably and it is an important feature of the Project Review and the Concept Hazard Analysis Review to determine when Safety Reviews are carried out. These not only are affected by the phases of the project: conceptual engineering, basic engineering, detailed engineering, procurement and construction, and commissioning but by the timing of capital authorisation and safety reports if needed. The need for reviews is also affected by the knowledge available about the project and the perceived hazards and risks presented by the plant.

Certain projects will be on a fast-track and the project phases will be merged. This also applies to modifications. Safety reviews of existing plant should be as appropriate to the experience and knowledge built up on the plant. An adapted form of Preliminary Hazard Analysis may be preferable to a HAZOP study.

Table 2: The Timing of Project Reviews

TIME
(not to
scale)

CONCEPTUAL	**ENGINEERING**

Project Review

Concept Hazard Analysis: Initial Review

BASIC	**ENGINEERING**

Concept Hazard Analysis

Critical Examination
Preliminary Consequence Analysis
Process Safety Sociotechnical System Review
Preliminary Hazard Analysis

Task Analysis/HAZOP (Batch)

DETAILED	**ENGINEERING**

Hazard and Operability study
Task Analysis, FMEA (if required)
Safety Report
Process Safety Management Audit
Environmental Audit

PROCUREMENT &	**CONSTRUCTION**

Process Safety Management Audit

Operability Review
Precommissioning Review
Precommissioning Inspection

COMMIS	**SIONING**

BENEFICIAL	**PRODUCTION**

Operating Review

Process Safety Audit (periodic)

Incident Reports (as they occur)

HAZOP (periodic)
Process Hazard Review (periodic)
Special studies as determined

135

The Hazard Analysis should be reviewed and updated periodically with typical review intervals ranging between 3 and 10 years. The Hazard Analysis for existing facilities might be considered in the following priority[10]:

a. High Substance Hazard Index value or large quantities of toxic, flammable or explosive substances.

b. Proximity to a populous area or a plant location where large numbers of workers are present.

c. Process complexity, including strongly exothermic reactions.

d. Severe operating conditions such as high temperatures or pressures or conditions that cause severe corrosion or erosion.

It should not be forgotten that critical activities are determined by production, quality and safety factors and studies can be activated by any of these.

Overruling considerations include an incident, changes in technology or in the facility. Legal factors can enforce change. Also the experienced gut reaction of the Safety Officer and others can be an important factor.

Conventionally such a study might involve either the update of a HAZOP or a retro-HAZOP for plants which predate HAZOP studies. For such plants there is considerable information on the plant history, its operation and maintenance, data on failures and incidents, etc. Consequently modified procedures such as Preliminary Hazard Analysis and Consequence Analysis can be used to reduce the effort involved.

5. Risk Assessment Methods

Risk assessment is the complex process used to describe the study of decisions subject to uncertain consequences and may be subdivided into risk estimation and risk evaluation. The latter involves determining the significance or value of the identified hazards and estimated risks to those concerned with or affected by the decision. Risk analysis involves making value judgements based on both accepted and disputed information.

Residual risk is the risk remaining after all proposed improvements to the facility under study have been made. Residual risk is always present and its study is continued throughout the life of the plant to *at least* maintain the initial project standards and revise these standards as Company or Regulations require.

Risk Indices give an indication of deviations from good practice. Some checklists and hazard indices like the Dow and the Mond Index give an impression of the relative risk of a proposed plant. One of the best methods of getting an estimate of risk is from the inventory levels and conditions under which dangerous substances are stored. For convenience all such estimates will be termed Risk Indices.

Short-Cut Risk Assessment Method, SCRAM gives a semiquantitative estimate of relative risk by summing the likelihood of a specific top event with an arbitrary severity category. The fact that the severity categories are based on value judgements means that this is a risk assessment method.

Quantitative Risk Assessment, QRA is used to evaluate the likelihood of a specific incident and the extent of the consequences. It can be used to evaluate individual and societal risk. QRA normally involves the use of the following techniques:

- *Fault Tree Analysis, FTA,* for representing the logical combinations of various system states which lead to a particular outcome; an undesired event known as the top event. A fault tree model can only relate system states with difficulty and often stops at the release of material when modelling an incident scenario.

- *Event Tree Analysis, ETA,* follows a cause through to the possible outcomes, branching at each point when there is more than one possible result from the precursor event, until the final outcomes of interest are reached. It is widely used in consequence analysis and some variations are used in evaluating human actions.

- *Cause-Consequence Analysis, CCA,* also follows through causes to events but allows for the use of gates. It is necessarily more complicated than FTA and ETA and is not used widely.

Other techniques are being developed to evaluate risk but in general the Chemical Industry uses a homeostatic regulation process to achieve at least pre-set goals although it accepts that other forces will be changing the targets as a result of the collaborationist regulation process[6].

6. The use of Risk Assessment

The Health and Safety Policy of a company has to manage all activities so as to avoid causing any unnecessary or intolerable risk to the health and safety of all employees, customers and members of the public who may be affected by its operations. Similar aims affect property and the environment, with particularly tight constraints on risk for listed buildings and sites of special scientific interest. Such objectives require not only compliance with all legal requirements and appropriate Codes of Practice, but taking additional measures to eliminate hazards whenever reasonably practical, and reduce the average accidental risk for employees at work to no more that their average risk at home.

The process industries have found that such a policy necessitates that hazard studies are commenced well before detailed design in order to identify associated significant hazards and reduce the residual risks to acceptable levels by appropriate design and selection of process materials, equipment and operations. Subsequently vigilance must be maintained during production such that risks are kept at or below a tolerable level. It is also necessary to have installed an effective in-house Safety Management System available which after appropriate review to allow for the new plant will ensure that all organisation and management features are in place.

Performing a short-cut risk evaluation leads to a better understanding of the system - particularly of incident scenarios, hazard identification and human response in an emergency. Generating estimates of risk allows the benefit of risk reduction measures to be gauged. The method illustrated later uses estimates of likelihood and severity. The factors used in the severity categories encourage a complete study of the ways of causing damage and harm to property, business, people and the environment. The assessment values also serve to indicate where further work is required, including performing a detailed Quantified Risk Assessment.

In general the standards of safety in the process industries are high due to the implementation of rigorous management systems, the safety awareness of the workforce, and compliance with regulations, codes of practice and standards. QRA provides objective data to assist in judging divergent views on the allocation of resources to safety expenditure. It assists in making decisions as to whether to cease production on a large plant whilst repairs to a section are carried out. Of course there will still be disagreement about the potential hazard or the tolerability to risk of the public around the site. However some of the arguments can then be related to quantitative values which are capable of rigorous scrutiny.

Short cut risk evaluation should not be confined to project activities. It can also highlight the risk arising from maintenance activities or from continuing production when protective devices are down. The analysis can be built into the Safety Schedule of the plant which assists in the analysis and serves as a record of the design intent and the safety performance throughout the life of the plant..

7. Tolerable risk and uncertainties in estimates

The target of any company is to have zero accidents. However it is inevitable that unlikely major events will occur at some location because of the large number of companies worldwide. So arguments have been accepted by Regulators that the staff of an operating company should when carrying out a design use a company standard which sets target values for the maximum risk which might be tolerated from their activities.

The current company target values for land-based operations which companies in the UK appear to be using are similar to those given in Table 3.

The highest targets might be exceeded by an order of magnitude (x 10) in circumstances deemed important by the company. If any vulnerable groups were in the vicinity then the higher values would apply. These are compromise best estimates obtained from canvassing opinions in industry and values quoted by HSE. In the UK these values have been greatly influenced by publications on risk criteria for land use planning, HSE (1989)[7]. An intolerable risk is specified which cannot be justified on any grounds, say an individual risk of 10^{-4} fatalities per year; a broadly acceptable region is stated in which the risk is considered by the Regulators to be negligible, say 10^{-6} per year; and in between is the ALARP region where the risk should be as low as reasonably practicable and only undertaken if a benefit is desired.

Table 3. Target values of risk

Employee individual risk	
All process causes	3×10^{-5} per year
Specific process cause	10^{-5} per year
Public individual risk	
All process causes	10^{-5} per year
Specific process cause	10^{-6} per year
Risk of major incidents (i.e. societal risk)	
Near miss from all process causes	10^{-4} per year
Accident from all process causes	10^{-5} per year
Catastrophic accident from all process causes	10^{-6} per year
Accident from specific process causes	10^{-6} per year
Catastrophic accident, specific process causes	10^{-7} per year

Such target values would probably be considered acceptable by many social categories of people and someone working at the location. But they certainly would not be accepted by groups including a majority of egalitarians or sectarianists, see Royal Society Study Group (1992)[6]. Any member of the public would be aggrieved to find a chemical development in 'their backyard'. So the debate will continue and increasingly people have the right to know. This has increased the extent of disputes over risk values and marked the end of pronouncements from on high of acceptable risk...and quite right too given the uncertainties in the data.

An absolute estimate of risk is compared with specific target values of estimated risk. This is therefore highly sensitive to uncertainty resulting from errors in the evaluation due to incompleteness or inaccurate manipulation of data. Some of the sources of uncertainty in QRA are noted in Table 4. Only a small proportion of the incident scenarios which can occur can be identified in the time available so the emphasis is on successfully identifying the significant ones. Some immediate causes will certainly be missed although appropriate mitigation will be included. Where important causes are omitted then the subsequent analysis is likely to underestimate the risk.

Typically the likelihood of a given top event in QRA estimates has an absolute uncertainty of one or more orders of magnitude, that is a difference such as exists between 10^{-4} and 10^{-3}. Such an order of magnitude uncertainty in individual risk often corresponds to a much smaller uncertainty in physical location of the isorisk contour from a flammable event as many physical effects diminish rapidly with distance. This is not necessarily the case for toxic events which can reach surprising distances in both gaseous and liquid phases.

The *relative* use of risk estimates is less sensitive to error as the resulting risk estimates are subject to similar uncertainties, many of which will cancel out when evaluating the *change* in risk. It is therefore possible to estimate the reduction in risk achieved through the modification of a system with considerable accuracy, and only cases falling near or into an intolerable risk zone need to be prioritised for detailed study.

Other criteria also affect decisions as to the suitability of a project or whether to keep an activity operating. These criteria include:

Acceptability criteria for effluents, emissions, wastes and noise.

Access and egress to the site and location.

Acceptability and consultation requirements of outside bodies.

Impact on site of notifiable status.

Economic criteria for justifiable expenditure and consequential loss.

Availability of local expertise, skills and training.

Considerable experienced judgement is needed to interpret all the results and allow for the uncertainties inherent in the calculation. It is also on occasion necessary to convince others that the risk is tolerable and under control.Absolute values should be communicated with caution. What can be said is that if a risk analysis has been carried out with reasonable accuracy to a stated level of tolerable risk then the resulting plant will have been designed to the accepted standards within the industry. Should such standards change then revision should be made to reduce residual risk. If the plant as designed is not acceptable to others then an appropriate judgement must be made as to whether modification is justified

8. Short Cut Risk Assessment Method (SCRAM)

In the Short Cut Risk Assessment Method, SCRAM, see Wells[5], the risk is defined as the Likelihood, L, of a specific undesired event occurring within a given period or in particular circumstances. The likelihood measure is frequency (per year). The Severity, S, is a measure of the expected consequence of an incident outcome.

The Target Risk is defined by the equation:

$$\text{Target Risk} \quad = \log_{10}10^{L} + \log_{10}10^{S} = L + S$$

L is the exponent of likelihood as measured by frequency, a negative value. Normally this is measured in frequency per year.

S is a severity category iwhich for the units of kikelihood per year is defined in terms of numbers ranging from 1 minor to 5 catastrophic.

The target risk is only acceptable when its value is equal to, or less than, zero.

Table 4. Sources of Uncertainty in QRA

System description

Process description, drawings, or procedures do not represent actuality.

Site area maps and population data may be incorrect or out of date.

Available weather data may be inappropriate.

Hazard identification

Failure to identify all the significant failure events

Poor modelling of the incident scenario

Failure to include all significant events which have been identified

Identification of major hazards and their causes may be incomplete.

Hazard screening techniques may omit important cases.

Failure to incorporate all control measures in incident scenarios.

Frequency techniques

Extrapolation of historical data may overlook hazards from scale-up.

Limitation of fault tree theory requires system simplification.

Incompleteness in fault and event tree analysis.

Data may be inaccurate, incomplete, or inappropriate.

Inherent problems in ascertaining human factors.

Frequencies modified by different management and maintenance factors.

Consequence techniques

Inappropriate model selection and validation.

Incorrect physical basis for model and uncertainties in physical data.

Source terms selected incorrectly.

Uncertainties in damage effects.

Mitigating effects incorrectly applied.

Risk estimation

Assumptions to reduce the depth of treatment.

Restricted conditions of wind speed and stability.

The results can be used to prioritise risk studies. To reduce the risk either reduce the likelihood of occurrence, which is a measure of the expected probability or frequency of occurrence of an event, and/or ameliorate the severity of the consequences or its occurrence by appropriate measures.

The severity categories which are assigned to any incident scenario should be based on the highest level indicated by the category. The 'major' consequences correspond to the ranking of 'high' used as the highest level by many companies. The levels 'severe' and 'catastrophic' relate to very rare events occurring at most one in 100,000 years and are included to reflect that such incidents may occur on a worldwide basis each year.

The need for short cut risk evaluation in the earlier stages of design, and certainly to HAZOP, has led to an increased emphasis in Concept Design and Preliminary Safety Studies. These methods are designed to effectively evaluate the incident scenarios within the area of identification of emergency control, 'dangerous disturbances', and consequence analysis, but the analyst accepts that the identification of all immediate causes of incidents will not be achieved. This method of assessing risk is sufficiently accurate to justify its use to prioritise further studies of risk

A list of possible incidents are generated as top events together with an indication of their incident scenario which is initially recorded on a preliminary hazard analysis sheet This is subsequently turned into the safety schedule which is then further developed by detailed design amongst other purposes. Table 5 shows the type of schedule developed for a critical scenario, albeit in abridged form. Order of magnitude values are generated for average duties for both plant and human actions. The method probably reduces the frequency of improper selection of incidents for analysis but clearly this is at the expense of complete generation of the individual incident scenarios.

9. Data Adjustment

A failure rate is not an intrinsic and immutable property of a piece of equipment. Values vary due to factors such as the severity of the processing medium and the operating environment, the suitability for service, the maintenance strategies adopted and factors related to the data itself and the defined equipment boundary.

The quality of data for use in risk assessment is generally poor with much published data stemming from sources going back past the 80's, for examples see Canvey[8]. This has achieved a certain status quo as it is perceived as giving historically correct answers. However, much of the data does not distinguish for example between low recovery, fast and dangerous failures and high recovery, slow and safe failures in control systems or data. It does not allow for changes in the historical record, such as have affected the likelihood of BLEVE's and the time available for evacuation, due to the change in relevant technology, fire-fighting and plant layout over the last 20 years. In addition some data can only be very approximate due to the large number of assumptions and uncertainties in the model. However in general the normal data used produces results within guidelines for tolerable risk. This indicates the use of similar standards rather than achieving an absolute value of risk.

Table 5. Preliminary Hazard Analysis: Main Disturbance

IMMEDIATE CAUSES	HAZARDOUS DISTURBANCE	INADEQUATE CONTROL OR ACTION	INADEQUATE EMERGENCY CONTROL / ACTION	DANGEROUS DISTURBANCE	FAILURE TO RECOVER THE SITUATION	SIGNIFICANT EVENT (RELEASE)
High inlet temperature to reactor F = 0.1	HIGH TEMPERATURE IN REACTOR F = 0.1	Operator fails to stop trend on TAH by adjusting control (slow) P = 0.01	High temperature shutdown system fails P = 0.05	OVER TEMPERATURE IN REACTOR F = 0.0005	Operator fails to stop all flows into Methanator P = 0.1	RUPTURE DUE TO OVER-TEMPERATURE F = 0.00005
High CO_2 in stream from absorber F = 0.1	HIGH TEMPERATURE IN REACTOR F = 0.1	Operator fails to stop trend on QHA(COx) or TAH by correcting absorber P = 0.1	High temperature shutdown system fails P = 0.05	OVER TEMPERATURE IN REACTOR F = 0.0005	Operator fails to stop all flows into Methanator P = 0.5	RUPTURE DUE TO OVER-TEMPERATURE F = 0.0003
Impurities in feed down start-up line (sneak path) F = 0.01	HIGH TEMPERATURE IN REACTOR F = 0.01	Operator fails to stop trend on QHA(COx) or TAH by correcting isolation P = 0.5	High temperature shutdown system fails P = 0.05	OVER TEMPERATURE IN REACTOR F = 0.0003	Operator fails to stop all flows into Methanator P = 0.5	RUPTURE DUE TO OVER-TEMPERATURE F = 0.0001

F = Frequency/year
P = Probability

A - Immediate attention needed
B - Further study probably required
C - Further study may be necessary

MAJOR RELEASE ON RUPTURE DUE TO OVER TEMPERATURE	FAILURE OF MITIGATION	FAILURE TO AVOID ESCALATION
F = 0.0004	P = 1	P = 0.01
Likelihood as power of frequency	3/4	5/6
Severity as category	3	4
Prioritisation for further study	C	C

Data must be adjusted by a range of organisational and management factors. The need to adjust all data according to these factors is an impracticable task, infeasible at an early stage of design. Fortunately for many processes the factors affecting consequences are not critical as the extent of possible damage is restricted and well defined. Such plants would normally only pose severe or catastrophic consequences given extremely rare events. It is also not difficult to distinguish between a soundly run works operating technology which is well understood by the work force and one badly maintained, having new or novel technology outside the skills of the work force. A process unit having novel technology at a site close to a populated area, but remote from technological support is obviously going to be more at risk.

The approach recommended is to look for major variances in one or more factors and change values by up to an order of magnitude. Typical changes to make between excellent, base case and worst case values are indicated in Table 6. If such adjustments affect key variables, such as the likelihood of immediate cause and the probability of inadequate emergency control then the risk is almost certainly going to appear unacceptable. This will then direct appropriate remedial action to either eliminate the deficiency or reduce its effects.

Table 6. Adjustment of average case values

Determine whether the plant conditions represent:

  a) Average duty or appropriate baseline failure rate data.

  b) Excellent case conditions for a clean internal duty with good maintenance performance is good on a well established plant

  c) Worst case conditions for dirty internal duty, poor maintenance performance and plant of novel technology

Adjust the likelihood of failure in average duties as follows:

  a) In the worst case multiply by 10

  b) In an average case multiply by 1

  c) In an excellent case multiply by 0.5

Modify the ineffectiveness probabilities of control and mitigation

  a) If a protective system is in a failed state, P = 1

  b) In the worst case increase values as follows:
      0.01 to 0.1; 0.1 to 0.5; 0.5 to 0.9

  c) In an average case do not adjust probabilities

  d) In an excellent case reduce values as follows:
      0.001 no change; 0.01 to 0.005, 0.1 to 0.05; 0.5 to 0.1

Table 7 indicates the adjustment of likelihood's from Table 6. These results may seem excessive but certainly correspond with performance in the early years of the plant's life. A design system comfortably within the company's tolerable risk criteria was unacceptable. In reality even worse values can arise: remember Bhopal[9].

Table 7. The effect of maximum adjustments

	Average case	Worst case
Immediate causes of deviation	F = 0.1/year	F = 1/year
Failure to control the situation	P = 0.1	P = 0.5
Failure of emergency control	P = 0.05	P = 0.5
Failure to recover situation	P = 0.5	P = 0.9
Significant release of material	F = 0.0003/year	F = 0.2/year
Failure of mitigation	P = 0.1	P = 0.5/year
Severe accident	F = 0.00003/year	F = 0.1/year

Quantified Risk Assessment is valuable at all stages in the life of a plant. All plants contain residual risk. Action to evaluate and reduce risk should continue throughout the life of a plant by seeking to eliminate the root cause of incidents - particularly with respect to maintenance, external threats, procedures, information, information transfer and information processing, the abilities of personnel in the task, and the capabilities of management and organisation.

10. References

1. AIChemE, 1989, *Guidelines for Technical Management of Chemical Process Safety*, Centre for Chemical Process Safety of the American Institute of Chemical Engineers, New York

2. Wells, G.L. Phang, C., Wardman, M.J. and Whetton, C.W. 1992, *Incident scenarios: Their identification and evaluation.* Process Safety and Environmental Protection. Trans. IChemE, Vol. 70, Part B, November. pp 179-188

3. Turney, R.D., 1990, *Designing plants for 1990 and beyond: Procedures for the Control of Safety, Health and Environmental Hazards in the design of chemical plant*, Process Safety & Environmental Protection, Trans. IChemE, Feb. p12-15.

4. Wells, G.L., Wardman, M.J. and Whetton, C.W. 1993, *Preliminary safety analysis.* Journal of Loss Prev. Process Ind., Vol.6, No.1, p47-60.

5. Allum, S. and Wells G.L. 1993, *Short-cut risk assessment.* Trans IChemE (in press).

6. Royal Society Study Group, 1992, *Risk: Analysis, Perception and Management*, The Royal Society, London.

7. HSE, 1989, *Risk Criteria for Land Use Planning in the vicinity of Major Hazards*, HMSO, London

8. HSE, 1978, '*Canvey - An Investigation of Potential Hazards from the Operations of the Canvey Island / Thurrock Area*', HMSO, London.

9. Bowonder, B. and Miyake, T. 1988, *Managing hazardous facilities: Lessons from the Bhopal accident.* Journal of Hazardous Material, Vol.19, No.3. p237 - 269

10. API, Jan. 1990, *API Recommended Practice 750*, American Petroleum Institute

Extending Safety Analysis Techniques with Formal Semantics

Janusz Górski

Franco-Polish School of New Information and Communication Technologies

60-854 Poznań, Poland

Abstract

Among the causes of many of the problems with safety analysis are impreciseness and ambiguity of the output data delivered by the safety analysis techniques and the resulting difficulties with interpretation of those data. An approach which can be undertaken to mitigate this problem is by providing the safety analysis techniques with more formal semantics. This paper aims to investigate this approach in more detail. First we give an overview of present practices during safety analysis. Then some problems with interpretation of the output from the presented methods are identified. This leads to the motivation to resolve ambiguities by adding more formality to the considered methods. The benefits of such approach are demonstrated by applying the formalism to some examples.

1. Introduction

In the present state of technology there is no single approach which can be used to analyse and assess safety. Instead, safety is assessed by developing a complex argument which combines results of many techniques and methods covering different aspects of the system under consideration and involving different teams of people over a considerable time scale. In current practice safety analysis suffers several deficiencies. With respect to system definition and description, often relevant subsystems or activities are excluded and the description does not correspond to the real world. With respect to hazard identification and accident modelling, important accident contributors or families of them are excluded or omitted. With respect to quantification of risks, component failure rates or human error data are uncertain and consequence modelling is inaccurate. With respect to documentation of results, boundaries and assumptions of analysis are not explicitly described and sources of quantitative data are not present. Human, organisational and training factors are among the most difficult items in safety analysis.

It seems that among the causes of many of the problems with safety analysis are impreciseness and ambiguity of the output data delivered by the safety analysis techniques and the resulting difficulties with interpretation of those data. An approach which can be undertaken to mitigate this problem is by providing the data with more formal semantics. This paper aims to investigate this approach in more detail. The paper is structured as follows. First we give an overview of present practices during safety analysis. Then some problems with interpretation of the output from the presented methods are identified. This leads to the motivation to resolve ambiguities by adding more formality to the considered methods. The benefits of such approach are demonstrated by applying the formalism to some examples.

2. Safety Analysis Techniques

In this section we give a brief description of three commonly used methods of safety analysis: Fault Tree Analysis, Event Tree Analysis and Failure Mode and Effect Analysis. Those are the methods which, in our opinion, can mostly benefit from being extended with more formal semantics. We do not cover here Hazard and Operability Study (HAZOP) which is a "structured brainstorm" - type method with the main stress on managerial aspects. However, as HAZOP may make use of FTA, ETA and/or FMEA, it can also benefit from the proposed approach.

2.1 Fault Tree Analysis

Fault Tree Analysis (FTA) is widely used in the context of safety applications [3,4]. The assumption behind the fault tree approach is that the failure space is easier to identify and describe than the success space (it is easier to agree on what is a failure than on what is a success). Also, the failure space is less structured - less failure classes or types are worth to be considered than it is a case from the success standpoint. It is also easier to sacrifice a part of the success space and to include it to the failure space e.g. in order to make the description of the failure space simpler. Because the failure space is less structured, there are usually few system failure modes which determine the number of fault trees to be developed (as opposed to many success modes which would have to be considered).

A *fault tree (FT)* covers a particular failure mode given by the *top event*. Consequently, it does not cover the total failure space of the system. It includes those faults which contribute to a given failure mode. Usually, only those contributing faults which are most credible according to the analyst's assessment are included. The basic semantic notion of FT is that of *event*. The following classification of events is applied.
- *top event* - an event representing the failure mode under consideration,
- *primary events* - events which causes are not considered. A primary event can be further classified as:
  - *basic event* - the event on the lowest level of resolution (by assumption, it will not be developed further down),
  - *undeveloped event* - temporarily not developed (it has insufficient consequence or information for its development is not obtainable),
  - *external event* - an event which is not a fault - it is normally expected to occur.

- *intermediate events* - fault events which occur between the top and the primary events.

FT is entirely expressed in terms of events and their relationships. Two basic types of relationships between events are represented.

OR gate

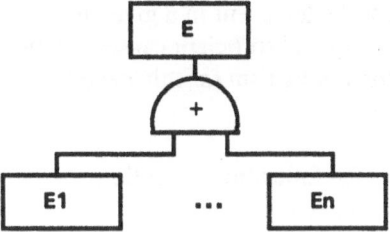

means "if E1 or..or En occurs then E occurs".

AND gate

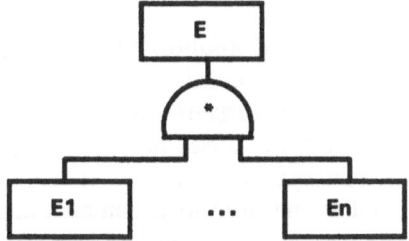

means "if E1 and .. and En occur then E occurs".

While considering the nature of failures it is useful to distinguish between the components of the system. *Passive components* are transmitters of "signals". The nature of "signals" may vary (force, current, fluid, energy, etc.). A passive component can fail by non transmitting or transmitting partially (no output or wrong output). Active components modify or originate signals, also changing their nature. They fail by providing no output, providing output without input or by providing wrong output. Quantitatively, passive components fail two-three orders of magnitude less frequently than active ones. The following are typical component failure categories:
- *primary* - is any failure of a component that occurs in an environment for which the component is qualified. In "software terms" this corresponds to the notion of incorrectness of the component, e.g. the component fails with input which satisfies the pre-condition,
- *secondary* - is any failure of a component that occurs in an environment for which it has not been qualified. E.g. the component fails because the input data are incorrect (pre-condition does not hold),

- *command* - a component functions properly but in wrong place or in wrong time. It involves *configuration faults*, e.g. the system is not linked properly, *usage faults*, e.g. the output is not used as intended, *timing faults* - the component is invoked in wrong time.

The FTA methodology comprises the following steps
- choose the system boundaries,
- identify possible failure modes of the system,
- for each failure mode build a FT: for the top event in a given FT
  - determine the immediate, necessary and sufficient causes for the occurrence of this event (find the mechanism for this mode),
  - continue this way down the tree.

In the FTA terminology, events can have the following semantics:
- a named change of state, e.g. "ignition key is turned",
- a state change which happened and continues to stay, e.g. "loss of 100 lives" (it can not be reversed),
- a sequence of events combined with some state conditions.

2.2 Event Trees

The following concepts are central to *Event Tree Analysis* [2]. *Safety functions* define what should be done in order to keep the system safe. They can be structured based on their logical dependencies, e.g." if the Reactor Coolant System (RCS) inventory is not maintained then RCS heat removal can not be accomplished". *Safety function event tree* represent temporal ordering of safety functions (depending on their failure or success) and the result of their common application. Thus, function event trees represent possible scenarios of safety function application in the case of an *initiating event*. *Accidents* can be classified by safety functions (an accident belong to a class if it occurs if a given safety function fails). Initiating events are those events which may lead to an accident. They represent groups of events with similar characteristics rather than individual events. Initiating events are candidates for being roots in fault trees (to investigate their causes). More fine classification (of initiating events) is by *mitigating systems* - because the same safety function can be fulfilled by diverse subsystems, and some initiating events belonging to the same class (according to by-safety-function classification) may make some of the mitigating subsystems unavailable whereas other events from the same class do not. Consequently the first group forms a subclass, because the safety function can be realised in limited ways comparing to the second class.

Event Trees and *Event Sequence Diagrams* are structures that depict the current state of the analyst's knowledge about function and system dependencies. *Event Sequence Analysis* is used to represent all possible success paths for initiating events. This can be done on much more detailed level than function event trees. In particular, different possibilities of achieving the same safety function (e.g. by diverse subsystems) are considered here. The alternative ways can be chosen depending on the plant and control system state (e.g. which parts are operable, what are values of the critical parameters, etc.). Event sequence analyses and documents an expected plant response to each initiating event. Event sequence diagrams are used to define the ordering of headings of system event trees. The heading represents the order of layers of the system tree, from root to leaves, each layer corresponds to a safety subsystem (fulfilling or failing to fulfil a given safety function, or its part).

System event trees are built by classifying subsystems into unique groups that can perform a given safety function. Each group has its own system event tree (which shows how the function is accomplished by this set of subsystems). System event trees may be richer than safety function event trees because they may involve the impact from the system side (which is not included in safety function definition) or may consider faults of multiple safety functions (due to some system faults). The ordering of events in the system event tree reflects the temporal, functional and hardware dependencies of subsystems the event refers to. Event trees contain success/fault events. In event trees, function (component) failure means that the function has not been performed as specified, or was performed at wrong time or was not performed at all.

Basic graphical structure for event trees is shown below. From each node of a tree which represents an activation of a given safety subject S, emerge two possible events: $\uparrow$ S (for success) and $\downarrow$ S (for failure).

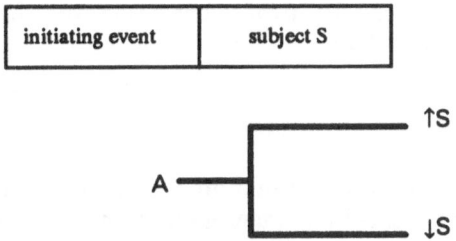

The occurrences in time of $\uparrow$ S and $\downarrow$ S are subjected to obvious restrictions, namely success excludes failure and vice versa.

2.3 Failure Mode and Effect Analysis

Failure Mode and Effect Analysis (FMEA) is intended to identify failures which have significant consequences affecting the system performance in the application considered [1]. FMEA assumes that the system structure has been identified down to the level where the *primary failure modes* are available. From this level, FMEA determines *secondary failure modes* which may occur on the higher levels of the structure hierarchy. FMEA can be applied in a limited way during conception, planning and definition phases and more fully in the design and development phases. FMEA assumes that the system structure is known (otherwise the failure consequences could not be identified). The development of block diagrams and state diagrams derived from the system structure is interrelated with FMEA. Basic semantic notions of FMEA are as follows:

- the system breakdown into components,
- diagrams of the system functional structure and identification of the various data which are needed to perform FMEA,
- the failure mode concept,
- the criticality concept.

The lowest level of interest is identified (component, functional) and the possible failure modes for each item at this level are listed. Those failure modes imply failure modes at the next higher level, etc. FMEA takes into account various modes of operation, e.g. system initiation, operation, control, maintenance. Because FMEA is applied throughout subsequent phases of the life-cycle, the information increments gradually (i.e. the relevant model must be able to represent an incomplete information).

FMEA may be conducted on hardware, system functions, may also include software and people. Common Mode Failures (CMF) are covered by FMEA. The causes for CMF-s can be classified into five main categories:

- environmental effects (normal, abnormal, accidental),
- design deficiencies,
- manufacturing defects,
- assembly errors,
- human errors (during operation and/or maintenance).

FMEA is an inductive method. It can be used alone. However, its natural complement is a deductive method (e.g. FTA)

3. Interpretation Problems

All three methods presented in the previous section are based on a rather rough notion of events and their relationships. An important issue is mutual influence of events. Should we treat events as atomic, instantaneous entities or should we allow that an event has a duration in time. In the first case, the consequence is that events can be totally ordered in time: if events are atomic (i.e. they can not coincide in time) then they must have happened in some order (which may be unknown at present). Note that the word 'time' is used here to represent evolution (cummulation of changes) of the world under consideration, not the physical time. In other words, an event can be treated as being instantaneous even if it has a duration in physical time. However, in this case the change of state represented by the event must be properly isolated from the other changes. The assumption about atomicity underlies most of the approaches to formalization of the event concept. Then, those 'events' for which this assumption can not be justified are represented as structures of more primitive, atomic events. Usually, the words 'action', 'transaction', 'operation', 'composed action', 'process', etc. are reserved to denote such events. Another important issue is the relation between events and the real time. Real time is frequently referred to during hazard analysis. Informal specification of events includes specification of event start time, event duration time as well as specification of time distances among events. A commonly accepted model for system behaviour is a set of (possibly infinite) state sequences. Thus, the specification techniques have two possibilities available to encode system specific information: encoding it within the state and encoding it as a restriction on the sequence (history) of states.

For instance, there is little said about the temporal dependencies of the input events of a fault tree gate. In basic interpretation of an OR gate there is no temporal dependency and hence no causal dependency between E1,..,En and E. Thus, the OR gate represents a *generalisation* of input events: an occurrence of any of input events E1,..,En means the occurrence of E. In such interpretation OR gates lead down through abstraction levels of a tree - provide for identification of different classes of a given fault [4]. However, in different interpretation [6] an OR gate can represent causality between input and output events. In this case any combination of input events causes the output event to happen. And again, this leaves room for different interpretations of temporal relations among input events and their relation to the output event.

For AND gates, there is a causality relation and hence the temporal precedence between the events E1,..,En and E. AND gates are the basic means to represent causality among events within the FT structure. However, usually it is left unspecified what does it mean that "the input events occur together". Some specific sequence dependencies are captured by "higher level" gates like the INHIBIT gate which allows to show the condition which has to be satisfied in order to input event Ei cause event E (it can be covered by AND with showing the order of events and assuming that the conditioning event continues till the input event happens). Another example is *priority* AND gate which is logically equivalent to an AND gate, with the added condition that the events must occur in a specific order [5]. Some fault trees distinguish *reparable* and *non reparable* faults. A non reparable fault, if it occurs, continues to exist forever. The reparable fault exists for a finite time period.

From the above we can see that events as used in safety analysis can be subjected to different interpretations. The precise meaning of what is an event and the precise meaning of event interrelationships are left unspecified. Moreover, the FTA methodology advises that there should not be any formalization in mind while a fault tree is being constructed (compare the recommendation from [4, p.II-4]). The motivation behind this is that the analyst should not be restricted by any formalism while identifying possible ways of event interaction. However, it is not our intention to formalise the process of fault tree construction. Instead, we advocate introducing more formalism into the meaning of already existing fault trees. Such a formalization should be accomplished after the tree has been constructed. The usual benefits attributed to formal specifications can be expected here, including unambiguous communication between people, precise and clear statement of failure scenarios and support for mechanical analysis.

3.1 Seeking Solution

It is felt that many of the interpretation problems illustrated above could be avoided or mitigated if the results of application of the safety analysis methods were given more formal semantics expressed in terms of a common formal model. The benefits of such step could be threefold:

- formal semantics would enable recording safety related information in a precise, unambiguous way;
- common formal semantics base would provide for interrelating data structures resulting from different analyses and this way obtaining a fuller picture of the safety related issues of the whole system;
- formal semantics would provide for application of mechanical analysis of the safety related data, which has a potential for being faster, more accurate and more reliable than paper-and-pencil work.

The idea of a *Common Safety Description Model (CSDM)* was originated during the author's work in the Safety Engineer's Workbench project under the Eureka program [7,8]. Some preliminary results were reported in [9]. The main idea is to treat CSDM as a sort of "glue" which sticks together a number of existing safety analysis methods which are already in use and are identified (by engineers) as relevant and useful.

The basic concept underlying hazard analysis methods is that of *event*. The following are the main relationships among events which are of interest:

- *causality* - two kinds of dependencies are of interest:
  - events which are the *necessary* prerequisites of event E to happen,
  - events which are *sufficient* for event E to happen.
  Note the difference between the two cases. In the former case we identify events which give E a chance to happen, whereas in the latter case we identify events which cause E to happen.
- *temporal* ordering (an event happens before another event, two events overlap, etc.),
- *generalisation* (an event denotes one or more of events from a given set),
- relation to *physical* structure (an event is associated with a physical component),
- relation to *logical* structure (an event is associated with a function),
- relation to *flows* inside the system (an event is associated with a flow of, e.g. materials, current, fluids, etc.),

Additionally, an important role in hazard analysis plays the classification of events into negative (*faults*) and positive ones.

The part of the physical world which is relevant for a given application problem always have some structure, i.e. is composed of identifiable components which interact among themselves. The components usually form a hierarchical structure based on the "is-part-of" relationship which provides for multilevel viewing, where the lower levels provide more details about what a given component is made of. Another structure which not necessarily coincides with the physical one is the functional structure of the application. In general, functions are the named properties of the application. How they are specified depends on the assumed model and its associated language. Among the possibilities are: input-output data transformations, data abstractions, processes with input and output ports, objects and operations, assertions on the system state and/or history. Again, functions are usually structured and the notion of a function being a sub-function of other function should have a precise meaning.

From the safety analysis point of view, functions can be classified according to their (safety) criticality. An initial criticality classification should be then extended by applying some rules which derive the criticality classification from the component structure (component-subcomponent relation), function structure (function-subfunction relation) and implementation structure (function-implemented-by-components relation).

The crucial concept of safety analysis is that of *failure*. Usually it is not enough to say that a failure is anything what is not legal according to some specification (e.g. a function does different transformation than it is said in its specification). We are interested in failure classes (modes) and want to have them defined explicitly. Failures must be attributable to functions as well as to components.

4. An Approach To Formalization

In this section we present an approach to our attempted formalisation. In the centre of the safety analysis ontology is the notion of *event* which denotes a change (or lack of the required change) of the considered piece of world. In fact, the whole effort of safety analysis is targeted at event identification and identification of their interrelations. We will assume the linear time model - we treat time as an infinite set of time moments, linearly ordered. Thus, time can be modelled by e.g. the set of real numbers. We will assume the existence of *events* which may occur in time. The same event can have many occurrences - this is distinguished by giving each occurrence its unique *label*. We distinguish between two classes of happenings: *timed events* (*actions*) and *instantaneous events* (*transitions*). Transitions take no time, i.e. their occurrence in time is specified by specifying a single time moment. An occurrence of an action x is specified by specifying two associated transitions: x_s and x_e which denote the start and the end of x. We assume the discrete event model, i.e. the number of events which occur between two given events is finite. One of the basic relations among events which is sought for during safety analysis is *causality*. It tells how events contribute to occurrences of other events. The dependencies between accidents and their originating events, which may occur far away (in time and in space), are identified and recorded by means of the causality relation. Another important concern of safety analysis is *nondeterminism*.

Identifying nondeterminism we learn about possible alternative ways which lead (through causality) to the results which are of interest.

4.1 The model

Our model comprises the following:

E - a set of *events*, and its elements to be denoted by X, Y, Z.

L - a set of *labels* which are used to uniquely identify event occurrences, with individual labels denoted by l, n, m.

A - a set of *actions* - labelled events plus a distinguished *silent action* $\perp$,

$\quad$ **A** $= (\mathbf{L} \times \mathbf{E}) \cup \{\perp\}$; individual actions to be denoted by x, y, z.

T - a set of *transitions* with its elements denoted by W; a transition is essentially an action name with a subscript s or e; **T** is related to **A** in the following sense $\mathbf{T} = \{x_s | x \in \mathbf{A}\} \cup \{x_e | x \in \mathbf{A}\}$

$\quad$ intuitively, for an action x, x_s denotes the *start* transition and x_e denotes the *end* transition of x.

$\prec_c$ - a *causality* relation on $(\mathbf{T} \times \mathbf{T})$, a partial order that is irreflexive and transitive.

$=_c$ - a *causality equivalence* on $(\mathbf{T} \times \mathbf{T})$, representing transitions that have exactly the same causes; in particular, for any transition W, $(\perp_s \prec_c W \vee \perp_s =_c W)$ and $(W \prec_c \perp_e \vee W =_c \perp_e)$ hold.

$\Re$ - the set of *real numbers*, and its elements denoted by r; the notation $[r_1, r_2]$ denotes a closed interval of real numbers.

Time$(\in \mathbf{T} \rightarrow \Re)$ a partial function which assigns real time to transitions. *Time* can also be interpreted as a set of pairs $((l, X), r)$, where $r \in \Re, (l, X) \in \mathbf{T}$ and *Time*$((l, X)) = r$. Elements of *Time* we will denote by t, u. *Time* is restricted by $\prec_c$ and $=_c$ in the following sense, assuming $\{w, w'\} \in \mathbf{dom}$ *Time* (domain of the function *Time*), $w \prec_c w' \Rightarrow$ *Time*$(w) <$ *Time*(w') and $w =_c w' \Rightarrow$ *Time*$(w) =$ *Time*(w'), and furthermore we demand that $\perp_s \in \mathbf{dom}$ *Time* $\wedge \perp_e \in \mathbf{dom}$ *Time*.

Start$(\in \mathbf{A} \rightarrow \Re)$ a partial function that, given x, returns r if $(x_s, r) \in$ *Time*, otherwise *Start*(x) is undefined.

End$(\in \mathbf{A} \rightarrow \Re)$ a partial function that, given x, returns r if $(x_e, r) \in$ *Time*, otherwise *End*(x) is undefined.

We stipulate that, for any action x, $x_s \prec_c x_e$ and
$(x_s, r) \in$ *Time* $\Rightarrow \exists r' \cdot ((x_e, r') \in$ *Time* $\wedge (r < r'))$ and
$(x_e, r) \in$ *Time* $\Rightarrow \exists r' \cdot ((x_s, r') \in$ *Time* $\wedge (r > r'))$.
From the above we see that an action always has a duration in time and once started it always ends.

We proceed with definitions of *temporal ordering*, $\prec_t$, and *temporal equality*, $=_t$ relations $(\subseteq Time \times Time)$. Let $t = (w,r)$ and $t' = (w',r')$; we define
$t \prec_t t' \leftrightarrow w \prec_c w' \wedge r < r'$ and $t =_t t' \leftrightarrow w =_c w' \wedge r = r'$. Note that for those transitions which occur in *Time*, the ordering $\prec_t$ is implied by $\prec_c$ and similarly $=_t$ is implied by $=_c$

Now we come to a relation about actions and their starting and ending time moments - a relation $\approx (\subseteq \mathbf{A} \times (Time \times Time))$ which is defined as follows: for each $x \in \mathbf{A}$, $x \approx ((x_s,r),(x_e,r')) \leftrightarrow (x_s,r),(x_e,r') \in Time$

Causal relations for actions are defined as follows. Let x and y be actions, $x \approx (t_s,t_e)$ and $y \approx (u_s,u_e)$. The relations $\prec_i$ and $\prec_h$ on $\mathbf{A} \times \mathbf{A}$ are defined based on the $\prec_t$ between transitions of actions: $x \prec_i y \leftrightarrow t_e \prec u_s$ and $x \prec_h y \leftrightarrow t_s \prec u_s$. If $x \prec_i y$, we say that x is *interior causal* of y; if $x \prec_h y$, we say that x is *head causal* of y.

Observe that in the definitions above, $Start, End, \prec_t, =_t, \prec_i, \prec_h$ and $\approx$ are *Time* specific, whereas $\prec_c$ and $=_c$ do not depend on *Time*. Different *Time* functions represent different behaviour in our model. Intuitively, when *Time* is undefined for a transition w, we understand that w never occurs in the corresponding behaviour. Note that the range of the function *Time* falls in the interval $[Start(\perp), End(\perp)]$.

For notational convenience, we introduce the following abbreviations.

- predicate *occur*: $\mathbf{A} \to B$ is defined as $occur(x) \leftrightarrow (x_s \in \mathbf{dom}\,Time)$.
- function ϕ (of type $\mathbf{E} \to p(\mathbf{A})$, where $p(\mathbf{A})$ denotes the powerset of $\mathbf{A}$) applied to every event to obtain a set of actions such that
$\phi(X) = \{(X,l) | l \in \mathbf{L}\}$.
- predicate *overlap*: $\mathbf{A} \times \mathbf{A} \to B$ is defined as follows
$overlap(x,y) \leftrightarrow \exists r \cdot (Start(x) < r < End(x) \wedge Start(y) < r < End(y))$.
- predicate *any_equal*: $\mathbf{A} \times \mathbf{A} \times \mathbf{A} \to B$ is defined as

$any\_equal(z,x,y) \leftrightarrow (occur(x) \wedge \neg occur(y) \Rightarrow z_s =_c x_s \wedge z_e =_c x_e) \wedge$
$(occur(y) \wedge \neg occur(x) \Rightarrow z_s =_c y_s \wedge z_e =_c y_e) \wedge$
$(occur(x) \wedge occur(y) \Rightarrow (z_s =_c x_s \wedge z_e =_c x_e) \vee (z_s =_c y_s \wedge z_e =_c y_e))$

We assume that the argument list of *any_equal* can be extended and if so, the definition lists all possible choices of the participating event occurrences.

- predicate *any_cause*: $\mathbf{A} \times \mathbf{A} \times \mathbf{A} \to B$ is defined as the predicate *any_equal* with the only difference that the relation $=_c$ is replaced by $\prec_h$.
- function *max*: $\mathbf{T} \times \cdots \times \mathbf{T} \to \mathbf{T}$ returns a transition with the greatest *Time* value among its arguments (and similarly for the function *min*).

4.2 Interpretation

A fault tree is viewed as a set of events connected by AND and OR gates. In giving a formal semantics to the fault tree, we use *actions* to interpret the events in the tree and predicates to interpret the gates. The system being modelled is characterised by the choice of objects in the model: $\mathbf{E}, \mathbf{L}, \mathbf{A}, \prec_c, =_c$ with their restrictions and the properties of real numbers. Possible behaviours of the system are characterised by *Time* functions. The semantics of the fault tree is a set of boolean expressions which define a set of *Time* functions for the system.

A function M is used to interpret the gates of the fault tree; it accepts an intermediate node of the tree with the input events X, Y and the output event Z, and returns a boolean expression - the *characteristic predicate* of the gate. An example of M function application to a *causal AND* gate (*C) is given below :

$$M(AND(X, Y, Z)) \stackrel{def}{=}$$
$$\forall z \in \phi(Z) \cdot \exists x \in \phi(X), y \in \phi(Y) \cdot$$
$$(occur(z) \rightarrow occur(x) \wedge occur(y) \wedge (x \prec_{kz}) \wedge (y \prec_{kz}))$$

The quantifiers link action variables (denoting event occurrences) with events. For simplicity, in the sequel they will be omitted assuming that z denotes an instant of event Z, x an instant of event X, etc. Note that the "guard" $occur(z)$ is needed because, when z does not occur, the causal relation $\prec_k$ becomes undefined. Also note that we do not require that z must occur if x and y occur (i.e. not always, the occurrences of x and y cause occurrence of z). Such stronger interpretation is also possible, however it is not mandatory - the fault trees are constructed backwards, from consequences to their causes and the fundamental question they answer is: "assuming that a given top event occurs, what are the all possible scenarios which could cause it?".
Interpretations of other gates can be found in [9].

5 Examples

5.1 Formalization of a Fault Tree

In this section we present the interpretation of an example fault tree (Fig.1). The tree represents the main hazard associated with computer control of the batch nitration process, the decomposition [6]. The event names are given in capitals and are included in the nodes of the tree. The tree comprises four gates. GATE1 is of type *causal AND* (*C). GATE2 is of type *causal OR* (+C). GATE3 is of type *generalisation AND* (*G). GATE4 is of type *generalisation OR* (+G).

158

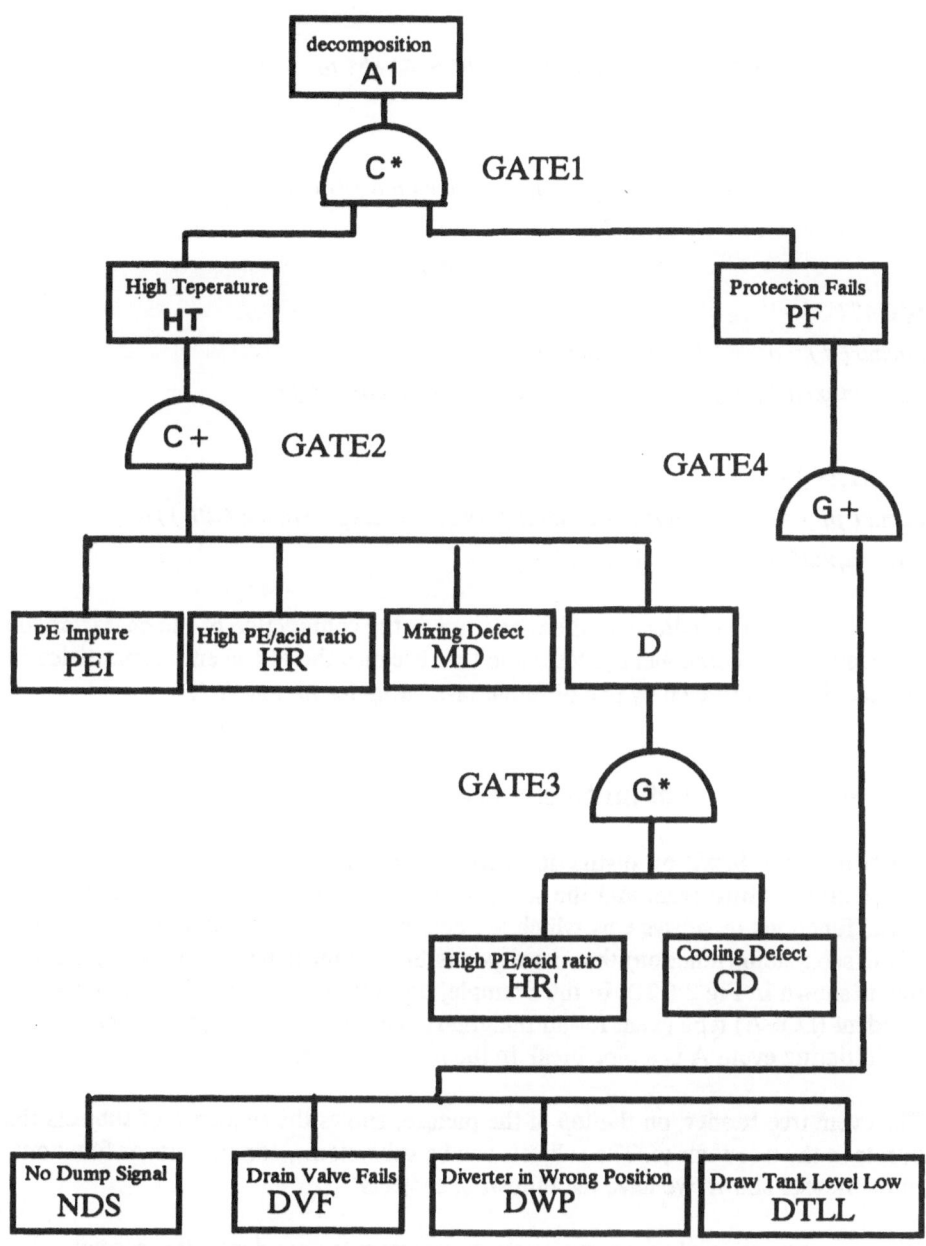

Fig.1. The fault tree for the batch nitration process decomposition.

$M(GATE1) =$

$(occur(a1) \Rightarrow occur(ht) \wedge occur(pf) \wedge (ht \prec _{M}a1) \wedge (pf \prec _{M}a1) \wedge overlap(ht,pf))$

$M(GATE2) =$

$occur(ht) \Rightarrow (occur(pei) \vee occur(hr) \vee occur(md) \vee occur(d)) \wedge$

$any\_cause(ht,pei,hr,md,d)$

$M(GATE3) =$

$(occur(d) \Rightarrow occur(hr') \wedge occur(cd) \wedge$

$(d_{s} =_{c} max(hr'_{s},cd_{s}) \wedge d_{e} =_{c} min(hr'_{e},cd_{e}) \wedge overlap(hr',cd))$

$M(GATE4) =$

$occur(pf) \Rightarrow (occur(nds) \vee occur(dvf) \vee occur(dwp) \vee occur(dtll)) \wedge$

$any\_equal(pf,nds,dvf,dwp,dtll)$

The characteristic predicate of the whole tree is the conjunction of the characteristic predicates of its component gates. Note that because the HR event occurs twice in the tree, it is interpreted by two different task variables hr and hr'.

5.2 Formalization of an Event Tree

Event trees are based on distinction between *success* and *failure*. The root of the tree is an initiating event and the subsequent layers down to the leaves correspond to the functions or subsystems which are activated in the case of success or failure of the subsystem (function) that corresponds to the higher layer. An example event tree is shown in Fig.2 ([2]). In the example, the initiating event is a loss-of-coolant accident (LOCA) type event for an imaginary nuclear reactor system. In particular, the initiating event A is a pipe break in the primary system.

The event tree header, on the top of the picture, shows the sequence of subjects the events in the tree refer to. The subjects can be subsystems, components or functions. In the header shown we have the following subjects:

B   - operation of the reactor protection system to shut down the reactor,
C   - injection of emergency coolant by pump 1,
D   - injection of emergency coolant by pump 2,
E   - post-accident decay-heat removal.

Subject activations are nodes in the tree. The branch leading up shows the success of the corresponding subject (i.e. the subject being the header of the column where the event is situated). The branch leading down represents the failure of the subject.

nitiating event	B	C	D	E	resulting event

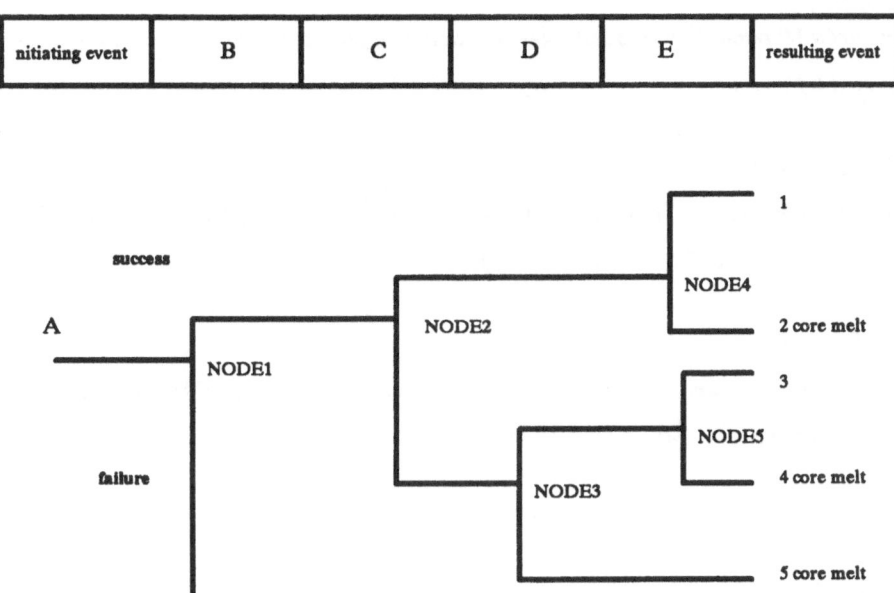

Fig.2. The event tree for a nuclear reactor system.

The six event sequences resulting from the tree are numbered on the right hand side of the picture. They are as follows:

1. A↑B↑C↑E - protection system successful, the pump 1 injects the emergency coolant and the heat is removed;
2. A↑B↑C↓E - protection system successful, the pump 1 injects the emergency coolant and the heat not removed - the core melts;
3. A↑B↓C↑D↑E - protection system successful, the pump 1 fails but the pump 2 works and the heat is removed;
4. A↑B↓C↑D↓E - protection system successful, the pump 1 fails but the pump 2 works and the heat is not removed - the core melts;
5. A↑B↓C↓D - protection system successful, the pump 1 fails and then the pump 2 fails too - the core melts;
6. A↓B - the protection system fails - the core melts.

The formalization of the tree is given below.

$A, \uparrow B, \downarrow B, \uparrow C, \uparrow C, \uparrow D, \downarrow D, \uparrow E, \downarrow E \in \mathbf{E}$,

$M(NODE1) =$

$(occur(a) \leftrightarrow occur(\uparrow b) \oplus occur(\downarrow b)) \wedge (occur(a) \wedge occur(\uparrow b) \Rightarrow (a \prec_h \uparrow b)) \wedge$

$(occur(a) \wedge occur(\downarrow b) \Rightarrow (a \prec_h \downarrow b))$

$M(NODE2) =$

$(occur(\uparrow b) \leftrightarrow occur(\uparrow c) \oplus occur(\downarrow c)) \wedge (occur(\uparrow b) \wedge occur(\uparrow c) \Rightarrow (\uparrow b \prec_i \uparrow c)) \wedge$

$(occur(\uparrow b) \wedge occur(\downarrow c) \Rightarrow (\uparrow b \prec_i \downarrow c))$

$M(NODE3) =$

$(occur(\downarrow c) \leftrightarrow occur(\uparrow d) \oplus occur(\downarrow d)) \wedge (occur(\downarrow c) \wedge occur(\uparrow d) \Rightarrow (\downarrow c \prec_h \uparrow d)) \wedge$

$(occur(\downarrow c) \wedge occur(\downarrow d) \Rightarrow (\downarrow c \prec_h \downarrow d))$

$M(NODE4) =$

$(occur(\uparrow c) \leftrightarrow occur(\uparrow e) \oplus occur(\downarrow e)) \wedge (occur(\uparrow c) \wedge occur(\uparrow e) \Rightarrow (\uparrow c \prec_i \uparrow e)) \wedge$

$(occur(\uparrow c) \wedge occur(\downarrow e) \Rightarrow (\uparrow c \prec_i \downarrow e))$

$M(NODE5) =$

$(occur(\uparrow d) \leftrightarrow occur(\uparrow e) \oplus occur(\downarrow e)) \wedge (occur(\uparrow d) \wedge occur(\uparrow e) \Rightarrow (\uparrow d \prec_h \uparrow e)) \wedge$

$(occur(\uparrow d) \wedge occur(\downarrow e) \Rightarrow (\uparrow d \prec_h \downarrow e))$

6. Conclusion and Future Work

The problem presented in this paper results from the observation that by providing the presently used safety analysis methods with formal semantics a number of ambiguities can be removed and a possibility of misinterpretation severely reduced. It is also expected that by integrating (through a common semantics base) a number of existing safety analysis techniques the combined effect of their application can well exceed the simple sum of their individual effects. The paper introduces the problem and presents some steps towards its solution. Some other results in this direction have been recently published by other outhors [10]. Much more work is necessary, however. The research will be continued in association with the SHIP (Safety of Hazardous Industrial Processes) project curried out under the Environment programme supported by the Commission of EC.

Acknowledgement

This work was supported by the SHIP Project (ref. EV5V 103) curried out within the framework of the CEC Environment Programme.

References

1. *Analysis techniques for system reliability - procedure for failure mode and effect analysis.* International   Electrotechnical Commission, IEC Standard, Publication 812, 1990.
2. *NUREG/CR-2300.*
3. Leveson N.G, Harvey P.R.: *Analysing software safety.* IEEE Trans. on Software Engineering, SE-9(5), 1983, pp. 569-579
4. *Fault Tree Handbook.* NUREG-0492, 1981.
5. Dugan J.B. and Bavuso S.J.: *Fault Trees and sequence dependencies.* Proc. Annual Reliability and maintainability Symposium, IEEE, 1990, pp. 286-293.
6. Health and Safety Executive: *Programmable Electronic Systems in safety Related Applications, General Technical Guidelines*, Her Majesty's Stationery office, London, 1987.
7. J. Gorski: *Towards a formal model of hazard analysis.* Techn. Rep. EUREKA SEW263, Sept. 1990.
8. E. Bloomfield, J. H. Cheng, J. Gorski: *Safety Analysis: a feasibility study into the development of a generic Safety Description Method (SDM).* EUREKA Project SEW 263, Adelard, 1990.
9. E. Bloomfield, J. H. Cheng, J. Gorski: *Towards a common safety description model.* SAFECOMP'91, Trondhaim, Norway, 1991.
10. G. Bruns and S. Anderson: *Validating Safety Models with Fault Trees.* Proc. SAFECOMP'93, (J. Gorski Ed.), Springer-Verlag 1993.

Social Issues in High-Tech Safety

Denis Jackson
Data Sciences UK Limited

Abstract

This paper discusses twelve social issues which, it is argued, already inhibit or will inhibit the development of reliable Safety-Critical systems. It is claimed that standards and regulatory systems have been prescribed with insufficient attention to some of these issues. Responsibility for remedial action is not always obvious, but it is hoped that exposure of the social issues will trigger further thought and some appropriate action.

1 Introduction

Over the past 10 years or more, the Safety-Critical community has concentrated on the technological issues of producing reliably safe systems. In so doing, it has left behind a wake of social issues which have barely been identified let alone resolved. However some issues are fundamental and threaten to impede progress in achieving safe systems unless they can be recognised, addressed and given appropriate action.

2 Recent History of Social Issues

Under the stimulus of a number of national disasters or accidents in the late 1980s, the Engineering Council produced an "Embryo Code of Practice" in 1991 [1]. After wide consultation, this was issued in 1992 as a "Code of Professional Practice" [2] and it addressed some of the issues raised in this paper. More recently, Margaret Tierney [3] has discussed sociological issues in the context of project management, and especially of work using Formal Methods, whilst Denis Jackson [4] and the Cooperative Bank [5] have reported on professional and business ethics respectively. There have been parallel activities in other countries, particularly in the USA by the ACM and IEEE.

The spate of standardisation activities of the last five years has also brought out into the open a number of latent social issues. For example, there is no consensus of opinion as to what are the objectives of having safe systems, and society has still to make up its mind as to whether the preservation of human life is to be supplemented by other preservation objectives. International and sector safety-critical standards are being drafted but appear uncoordinated on such fundamental issues and take disparate views of public accountability. The Ministry of Defence (MoD) has raised issues such as Safety Awareness [6] sub-clause 9a, [7] sub-clause 7.3.1 and Independent Assessment [6] clause 16. Professionalism and the ethics of "Fitness for Purpose" have been of concern to trade associations and well-established suppliers, in whose standards there are developing ideas concerning the qualifications and training of safety-critical practitioners.

3 Social Issues of Concern

3.1 Summary

Social issues which should be of concern to the Safety-Critical community are listed as:
  (a) Public Awareness.
  (b) Public accountability for safety.
  (c) Corporate safety awareness.
  (d) Independent assessment.
  (e) Ethics of "Fitness For Purpose".
  (f) Contracting.
  (g) Managing organisational conflict.
  (h) Handling semi-academics.
  (i) The management of complexity.
  (j) Acceptance of computer-generated "evidence".
  (k) Professionalism.
  (l) Sabotage.
The above issues are discussed in the next twelve sub-sections.
  Whilst it is not suggested that these are the only social issues of relevance, it is believed that they will inhibit the development of safety critical systems unless resolved.

3.2 Public Awareness

This issue has been addressed by the Engineering Council [1] [2] Section 10 and by Denis Jackson [4] Section 9.8. Public awareness is desirable because:
  (a) The public needs to be informed and able to take an objective view of the risks of safety-critical systems relative to more publicised risks such as those of road accidents, uncontrolled nuclear releases, air transport accidents etc.
  (b) Public perceptions ultimately result in legislation and regulation, which need to be pitched at a fair and reasonable level which is not simply reflecting transient topical emotions and the distortions of the media.
  (c) Safety-critical practitioners should be seen to be acting responsibly in the interests of the general public.
  (d) In the event of an accident, it is important to have a pre-planned crisis procedure which includes unambiguous communication with the public.
  (e) Young members of the public, and their parents, need to be convinced that there is a worthwhile career in safety-critical systems: otherwise recruitment will suffer and the existing situation of an inadequate, ageing and increasingly-expensive workforce will not be alleviated.
  The last point is of particular significance. There is a growing tendency for society to suffer from "acute accountability", e.g. to pillory professionals such as surgeons, doctors and other medical workers, who are only doing their best, when diagnosis or treatment goes wrong. After centuries of experience, the medical world is reasonably clear as to what constitutes good and bad practice, and the regulatory authorities can deliver peer judgements with confidence. However, the same is not true of the Safety-Critical, high-technology world. Here there is little consensus of opinion

165

concerning the detailed techniques of system design, production and assurance, and individuals are expected to make judgements, often without the support of true standards. This makes them vulnerable to retrospective accusations of malpractice. Professional Indemnity insurance is becoming very expensive and ultimately has to be paid for by society. Moreover, individuals will not relish having proceedings hanging over their heads, and their business in jeopardy, whilst difficult judgements are made by regulatory bodies or the courts. Unless we are careful, public demand will price itself out of safe systems, and the professionals will conclude that there are easier ways of earning a living.

3.3 Public Accountability for Safety

Organisations electing to use safety-critical systems have an obvious duty to protect their users, and they may take into account that the latter, voluntarily, accept a measure of risk in their occupation. However, "innocent bystanders" are a different matter, and now enjoy a right to redress, under the "strict liability" principle of the Consumer Protection Act (CPA) 1987, if they are injured due to the system. In addition, the interests of both users and the general public should be in the care of regulatory authorities, but identification of the latter can be difficult.

Under the Health and Safety at Work etc Act, 1974, the Health & Safety Executive (HSE) assumes overall responsibility for a number of sectors, such as the Nuclear, Railways and Offshore sectors which also maintain their own inspectorates. The Civil Aviation Authority (CAA) clearly regulates the air and ground aspects of UK civil aviation [8], and the Ministry of Defence regulates its own systems, but the MoD's internal delegation of regulatory powers is far from clear. According to [7] para 7.2, the responsibility appears to be that of the MoD (PE) Project Manager, who appoints an MoD Safety Assurance Authority to advise him.

Ideally, there should be a National index of Regulatory Authorities, of the procedures which they expect to be followed and of the standards which they set. The reality is very different: if the Regulatory Authority can be identified, his procedures and standards may be unclear and uncosted, so that the practicality and affordability of system certification may be in doubt. In fairness to regulatory authorities, it has to be said that they like to preserve a measure of absolute discretion over certification, without which discretionary powers some much-needed systems would never enter service, so that a measure of vagueness and flexibility is justified on pragmatic grounds.

Against this drab background of criticism, the UK Civil Aviation Authority (CAA) stands out as a good example for its regulation of avionic systems. Admittedly, civil aviation has been regulated as an international business sector since the formation of the International Civil Aviation Organisation (ICAO) in 1944, and the latter's lead and the maturity of its measures have benefitted the CAA. The CAA's corresponding regulatory measures for ground-based systems have yet to be promulgated.

To return to the Health & Safety at Work etc Act, even the issue of what constitutes a "safe" workplace is argued in the courts, especially whether a workplace is safe from reasonably foreseeable danger. In the context of Section 29(1) of the Factories Act 1961, it has been said that "safe" should be interpreted in its normal English dictionary sense, for to add gloss or technical meaning to the word would reduce the protection afforded to workers by the Act, and that there was no implication of reasonable foreseeability of unsafe aspects intended by the drafters.

Whilst on the subject of legal matters, it should be explained that four practical effects of the "strict liability" principle of the Consumer Protection Act 1987 are that:

(a) the source of a defect causing an accident is liable;

(b) no contract need exist between the liable party and the injured party;

(c) there is no requirement for the injured party to prove negligence on the part of whoever is liable;

(d) the provisions of the Act apply for a "limitation period", which can be for up to ten years from the date of supply, so that safety-critical systems having a long marketable life could still be sold many years after their design, and be a source of suppliers' liability for a further ten years.

As a result of the above, suppliers of general purpose products, such as processors, compilers, and network management software, should be concerned at their potential liabilities and lack of control over their risk. They cannot easily pick and choose their applications, which indeed may not initially be seen as safety-critical, but subsequent events could make them liable.

A complementary social issue erodes the effectiveness of the CPA 1987, i.e. how does an injured member of the public identify who is liable? The true cause may be obscured by a convoluted contracting and sub-contracting process which an outsider would find difficult to penetrate.

3.4   Corporate Safety Awareness

Many organisations are capable of supplying or operating complex high-technology systems, but when these systems are in the safety-critical area, one corporate characteristic which distinguishes the men from the boys is Safety Awareness. Unless this quality is present in supplier and operator, the system is at risk and possibly doomed to failure. Therefore prudent purchasers may increasingly consider it important to place contracts with firms which have an established cultural record of safety awareness, rather than being tempted by the attractions of cut-price quotations.

The Ministry of Defence (MoD) already requires that tenderers for safety-critical systems "provide reasonable evidence that there is an awareness of safety throughout their organisation" [7] para 7.3.1. So far as the author is aware, neither the MoD nor other major purchasers define "safety awareness" or what would constitute reasonable or acceptable evidence of that quality. Denis Jackson has suggested what might constitute reasonable evidence [9], and the revision of Def Stan 00-56 which was due for completion in late 1993 may clarify the MoD requirements.

Whatever the requirements, it seems clear that there is a general lack of safety awareness despite a few organisations being very professional and highly-tuned in this area. The DTI/Coopers and Lybrand market study [10] found that there was a significant lack of awareness amongst users of safety-critical systems. In this case, awareness was judged at the relatively coarse level of:

(a) knowledge of the existence of Safety-Related Computer Controlled Systems (SRCCS);

(b) knowledge of SRCCS standards;

(c) adequacy of information available on SRCCS;

(d) demand for general, as opposed to sector-specific, information on SRCCS;

(e) interest in joining an SRCCS club or association.

There was little comment on the awareness of suppliers, but general agreement was reported on the need for action in respect of awareness generation, information provision and standards. Just how awareness was to be generated was not specified.

The suggestion in [9] was that acceptable evidence of safety awareness might constitute:

(a) a policy statement, by the highest corporate authority, which asserts the organisation's motivation to supply (or operate) safe systems and to give high priority to that objective;

(b) the existence of a disciplined workplace and appreciation of its benefits to safety;

(c) the inclusion in corporate standards of appropriate safety margins such as engineering tolerances; computer processing power and storage; "safe subsets" of high level languages; defensive programming and the minimisation of complexity.

(d) recognition and understanding of the perceived accidental and malicious threats to the systems concerned, followed by pragmatic Hazard Analysis;

(e) evidence of the pervasiveness of safety consciousness within the organisation;

(f) documented data of past hazardous experiences.

To these points could be added closeness of the system developers to the users: in the software industry it has not been the general practice for programmers to work closely with purchasers or users, or even beyond their software managers or marketing interfaces with the purchasers.

3.5   Independent Assessment

Under this heading there are really two issues involved, i.e.

(a) What assessment techniques are appropriate?

(b) How independent should the assessors be of the producers of the system?

The former issue is well outside the scope of this paper in that there is no consensus of agreement on techniques at present and it would be fruitless here to enter into such discussion. However, it is pointed out that in the world of security assurance of information systems, there already exist four "Commercial Licenced Evaluation Facilities" (CLEFs) which have the necessary tools, personnel and other infrastructure for assessment. They are accustomed to assessing relatively compact security kernels of systems, although safety-critical systems are probably less compact. This then raises the social issue of whether society could afford the rigorous assessments of the CLEFs, even given the importance of Safety.

The independence required of assessors is currently being argued. MoD, in its original Interim Draft Def Stan 00-55, required total independence, but has since softened this requirement in [6] clause 16 and especially [7] sub-clause 7.7.3. The well-established suppliers of safety-critical systems, e.g. the avionics industry, have invested much time and money in pioneering their production and assessment systems (over the past 10-20 years) and do not see the need for change. They claim that their established systems embody adequate independence and are suspicious of the costs of casting aside their well-developed cultures, and of adopting new and idealistic practices. This paper supports the idea that the CLEFs could be developed to provide third party assessment, particularly as their independence would satisfy the most exacting regulatory authority, the code of confidentiality to which they work should satisfy the commercial interests of the system producers, and their charges should reflect the competitive environment in which they operate.

3.6   Ethics of "Fitness for Purpose"

A basic ethical issue is whether a system should be used, for the sake of its undoubted benefits, when it is known to contain undetected flaws, e.g. in software. Such systems range from automotive anti-lock braking systems to medical diagnostic systems, the latter sector and its ethics being discussed in [11].

Another basic issue concerns the fair distribution of scarce resources of skilled personnel and money to the production of safe systems. Against the background of the "classless society" being developed by UK Government, there is a risk that high-profile sectors or projects will receive resources to the detriment of other deserving causes.

Individuals may have to make decisions concerning, for example, whether to check an apparent requirement, or to improve a design, or to carry out additional testing, or to redesign as a result of testing. All these measures may be in the interests of safety but they cost money and may put a project "in the red" and bring accusations of "gilding the lily". As McFarland says of the US scene [11] "Computer professionals are well aware that too much loyalty to the public welfare could cost them their jobs and their careers. It has happened."

The Cooperative Bank's study of ethical attitudes [5] includes a survey of individual attitudes according to profession and position in the organisation. A broad conclusion was that bank managers were the most saintly with accountants at the other end of the spectrum. If the popular belief that accountants are rising to control many organisations is correct, it bodes ill for corporate ethical attitudes and safety.

3.7   Contracting

The ethics of purchasers are brought into question by their contracting practices. For example, purchasers may deny that the system which they are procuring is safety-critical, because of its marginal status and their wish to avoid the greater expense of a safety-critical system. A naive supplier, entering into such a contract, may find himself liable under the CPA 1987 in future years if someone is killed or injured by the system which he supplied.

A further example is provided by organisations which have a policy of minimum price contracting. This can result in the short-listing of unsuitable suppliers, and naive or unscrupulous purchasers and suppliers may contract together for a system which is destined to be disastrous.

3.8   Managing Organisational Conflict

Under this heading, Margaret Tierney [3] has identified problems of project management, the basic principles of which are historic rather than adapted to the scientific needs of safety-critical systems. The major project management priorities are likely to be concerned with meeting deadlines, controlling project expenditure and producing working systems, possibly resulting in inadequate system testing; poor documentation; use of sub-standard programmers; and systems being rushed into premature service.

Whilst such a picture is unlikely to apply to reputable suppliers of safety-critical systems, "project managers in mainstream commercial computing are not likely to allow quality considerations to override the adequacy of the project as a whole" [3] section 2.4. Confusions and acrimony will arise when the two communities meet in real projects as collaborators or in main/sub-contractor relationships.

3.9   Handling Semi-Academics

Safety-critical projects contain small pockets of specialised expertise, of which formal methods is a good example. Personnel having such expertise are needed only in small numbers and they need formal academic training and continued contact with academia. They can best be judged by their peers rather than conventional managers, and by their outputs, the latter needing to be comprehensible by normal mortals.

Therefore their management is a specialised corporate social matter which has been discussed by Margaret Tierney [3].

3.10   The Management of Complexity

By this term I mean the voluntary limitation of system complexity by the supplier and/or the purchaser. Modern systems increasingly involve component re-use, for which the arguments include superior "testedness". However, the re-used components are usually designed for wide appeal and in any particular application carry superfluous hardware or software elements. These elements can be a source of unreliability, yet it would be a brave man who removed them because this could make matters worse.

System complexity is at the heart of a vicious spiral of increasing system costs, increasing safety assurance costs, pressure on very limited skilled human resources, and pragmatic acceptance of reduced safety assurance. The social issue is that the public and purchasers should exercise restraint in the features which they demand of safety-critical systems, and purchasers should not put too much financial pressure on suppliers by "cut-price contracting" practices. In turn, suppliers should not take the easy way out in re-using components unless they have satisfied themselves that the safety case impairment is small or entirely justified. To design-out complexity requires a positive approach which is analogous to the dictum of the early aircraft designers to "simplicate and add lightness".

3.11   Acceptance of Computer-Generated "Evidence"

Society is certainly becoming heavily dependent upon computers and there is a tendency for society to accept as true anything which comes out of a computer. The UK courts are accepting digital storage and word processors as a necessary technology. However, there are much more advanced technologies whose potential use, or abuse by naive or unscrupulous persons, should give rise to safety-related social concerns.

For example, there can be few sectors more orientated to safety than aviation, in which computer technology is now available and used in order to:
(a) replay aviation accident sequences based upon the information available from "black box" flight recorders;

(b) simulate possible accident sequences where no actual records exist.

In the former activity, the data exists as samples taken, at typically 4-second intervals, of from 4 to 64 sensor outputs. Exceptionally, individual sensors could be sampled eight times a second. Computer technology is then used to recreate the flight by interpolating between these known observations to give an illusion of continuous observation of the accident sequence: impressive technology is then available to display, graphically, different views of the flight, e.g. a pilot's eye view, an external observer's view, an Air Traffic Controller's view etc. The technological ambiguities are due to fact that the sensor observations are not truly synoptic but could have taken place eight times a second or at any time within a scan period of, say, four seconds and that much can happen in aviation in such a time. Therefore to reconstruct the flight by assuming times for events and by assuming that parameters behave linearly between observations, is to play with fire. In aviation accidents, the fate of an aircraft can be determined by events and crew decisions made on shorter timescales than 4 seconds: when that sort of performance places unreasonable demands on human beings, automatic control functions are used, e.g. autopilots and full authority digital engine control units, and their operation cannot be reconstructed in the same way.

The second type of activity, simulation, has been developed to a high degree for flight training and aircraft design and research purposes. The fidelity of such simulators is accepted for those applications, even to the extent that a pilot's training for a particular type of aircraft may be conducted entirely in a simulator. However the use of simulators to suggest accident sequences may be helpful, but not necessarily authoritative.

The social issues are liable to arise if analyses from such sources are presented visually to, say, a judge and jury. The impact of modern graphics is considerable and could be exploited by unscrupulous and well-funded lawyers anxious to portray only views favourable to their case. Less affluent litigants, particularly those unable to call upon authoritative technical opinion to assess the "evidence" thus presented, may be unable to prevent a judge and jury gaining a false impression that truth is being presented. The courts are already suspicious of past forensic evidence presented to them in their less critical days and when they had not the benefit of modern technology: history could repeat itself unless the similar problem in relation to safety can be recognised now and routines developed for evaluating technical evidence.

A more general example is afforded by system "Safety Cases" which are becoming so complex (perhaps tens of thousands of pages) as to require computer power to compile, process, store and retrieve elements of the safety case. Such computer systems are likely to be large, complex and, for some time to come, experimental. The ASAM project [12] is probably unique in developing the first concepts of such a tool, but it does not appear to anticipate some social issues eg:

(a) when safety cases are developed in the public interest, how they should be stored securely in perpetuity;

(b) how contemporary safety case software tools and hardware will continue to be available to retrieve safety cases, especially as computer systems become obsolete in 5-10 years: many organisations must admit to having historic computer-based information which is no longer accessible;

(c) the scale of configuration control of the database and the need to be able to audit each stage of development and the data on which decisions were based;

(d) the system integrity, from the point of view of individual accountability for

171

actions;

(e) the durability of the storage media, be they magnetic, optical or other media yet to be made available;

(f) the legal acceptability of such data as evidence.

The scale of such social issues is proportional to the length of time during which safety case information may need to be retrieved: the "limitation period" inherent in the Consumer Protection Act 1987 indicates that a safety case needs to be retained until at least 10 years after the supply of the last system using that case.

3.12 Professionalism

Within the Safety-Critical community, professionalism is an evolving social issue with no organisation paramount in defining and setting standards. Ideas vary, and initiatives in this area have been taken by the Engineering Council [1], [2], the IEE [13], [14], the British Computer Society (BCS) [15] and the author [4], [16]. Unfortunately, only a small proportion of existing safety-critical practitioners have an allegiance to the professional bodies mentioned above.

My own view of professionalism in the Safety-Critical area is that it entails:

(a) constant updating of relevant technical knowledge by active participation in safety-critical projects, seminars, conferences etc. and extensive reading in the mass of papers, standards etc. currently being produced;

(b) the willingness, ability and motivation to gain any additional knowledge necessary for a particular task;

(c) the honesty to recognise one's own limitations and then to seek expert guidance to remedy deficiencies;

(d) the ability to make valid judgements and deductions from the relevant facts assembled concerning particular issues;

(e) having a position within a circle of associates sharing a common interest in reliable systems, and maintaining dialogues.

3.13 Sabotage

Hazard analysis of potentially safety-critical systems has evolved on the assumptions that hardware suffers from wear and tear in normal use, that logic contains accidental design errors made in good faith, and that all systems are subject to environmental occurrences. The notion of malicious interference or sabotage is not generally considered, but it puts a very different complexion on Hazard Analysis and there are many indications that modern society contains sick minds which could behave irresponsibly.

Such malicious interference is normally the parallel province of security, and in some sectors, such as the nuclear sector, safety and security are inseparable. However a threat of high-tech sabotage would generally be new, and isolated examples could stimulate multiple "copycat" incidents: new views would then have to be taken of the risks in existing systems.

4 Conclusion

It is concluded that safety-critical systems raise many social issues and that responsibility for remedial measures is not obvious. Individual practitioners are exhorted by professional codes of practice etc to behave ethically, but organisations rarely ascribe corporate responsibility for business ethics or address the sort of social issues which are raised in this paper. At higher levels of responsibility and planning, purchasers, trade associations, standards makers, regulatory authorities and Parliament itself are tentative in their leadership on social issues of safety-critical systems: this is probably due to sporadic public interest and to ignorance of potential costs and their affordability.

However some, if not all, of these social issues cannot just be set aside, because they are fundamental to the working practices, standards and organisational cultures necessary for the achievement of safe systems. It is hoped that by the exposure of some of the social issues in this paper, the disparate organisations capable of rectifying deficiencies will be stimulated to further thought and action.

References

1. The Engineering Council. Engineers and Risk Issues - An 'Embryo' Code of Practice. Engineering Council Discussion Document, 1991.
2. The Engineering Council. Engineers and Risk Issues - Code of Professional Practice, October 1992.
3. Tierney, M. Potential Difficulties in managing Safety-Critical Computing Projects: a Sociological View. Springer-Verlag London Limited 1993, 43-64. In Directions in Safety-Critical Systems, Ed. Felix Redmill and Tom Anderson, Safety-Critical Systems Club.
4. Jackson, Denis. Target Qualities of Safety Professionals. Chapman and Hall, London 1993, 9, 139-152. In Safety-Critical Systems - Current issues, techniques and standards, Ed. Felix Redmill and Tom Anderson, Safety-Critical Systems Club.
5. The Co-operative Bank. The Co-operative Bank Survey of Business Ethics in the UK. University of Westminster 1993.
6. Ministry of Defence. The Procurement of Safety Critical Software in Defence Equipment - Part 1: Requirements. Defence Standard 00-55 (Part 1)/Issue 1, 5th April 1991.
7. Ministry of Defence. Hazard Analysis and Safety Classification of the Computer and Programmable Electronic System Elements of Defence Equipment. Defence Standard 00-56/Issue 1, 5th April 1991.
8. UK Secretary of State for Transport. Civil Aviation Authority: Chairman's Objectives. Letter to the Chairman, CAA, 24th July 1986.
9. Jackson, Denis. Demonstrating Safety Consciousness. Safety Systems In: The Safety-Critical Systems Club Newsletter Volume 1, Number 3, May 1992.
10. Department of Trade and Industry. Safety Related Computer Controlled Systems Market Study. London: HMSO, Undated (but approx late 1992/early 1993).
11. McFarland, Michael C. Ethics and the safety of Computer Systems. Computer, February 1991, 72-75.

12. Forder, Justin, et. al.. SAM - A Tool to Support the Construction, Review and Evolution of Safety Arguments. Springer-Verlag London Ltd 1993, 195-216. In Directions in Safety-Critical Systems, Ed. Felix Redmill and Tom Anderson, Safety Critical Systems Club.
13. McGettrick, Andrew. The IEE draft policy on educational requirements for safety-critical systems engineers. Chapman and Hall, London 1993, 11, 160-166. In Safety-Critical Systems - Current issues, techniques and standards, Ed. Felix Redmill and Tom Anderson, Safety-Critical Systems Club.
14. The Institution of Electrical Engineers. Safety-related Systems - Professional Brief, September 1992.
15. Falla, Mike. Safety-critical software professionals in the BCS industry career structure. Chapman and Hall, London 1993, 14, 181-189. In Safety-Critical Systems - Current issues, techniques and standards, Ed. Felix Redmill and Tom Anderson, Safety-Critical Systems Club.
16. Jackson, Denis. New Developments in Quality Management as a Pre-requisite to Safety. Springer-Verlag London Limited 1993, 257-269. In Directions in Safety-Critical Systems, Ed. Felix Redmill and Tom Anderson, Safety-Critical Systems Club.

HUMAN ERROR IN THE SOFTWARE GENERATION PROCESS

Trevor Cockram Rolls-Royce plc
Jim Salter and Keith Mitchell Lloyd's Register
Judith Cooper and Brian Kinch Lucas Engineering & Systems
John May Open University Dept. Computing & Nuclear Electric

Summary

In this paper we discuss how faults are introduced into the software generation process. The Fault Analysis of the Software Generation Process (FASGEP) project has classified faults into Random and Symptomatic. Symptomatic faults are those faults where the input to the process was correct; but the output from the process was incorrect due to an error in the process. Random faults are those faults for which no specific cause for the fault can be identified, This paper discusses the nature of random faults and to what extent they can be attributed to human error.

Software process decomposition shows that the human engineering process can be described as a combination of intellectual (novel) processes and more mechanistic processes. The mechanistic processes, e.g. the use of tools, can be considered to be a specific form of human-computer interaction via a particular human-machine interface.

The classification of human error within the FASGEP project takes into account the work of Rasmussen, specifically the classification of Rule-Based errors, Skill-Based errors, and Knowledge-Based errors. The causal relationships for each of the classifications have been developed into a causal network. A type of graphical probability model is based on this.

A probability model, of which GPMs form a part has been generated to determine the susceptibility of a software generation process to the introduction of faults. The model uses evidence from metrics collected on team, management, environment and communication attributes.

1. Introduction

Software has been developed for use in many safety critical and safety related applications. In using software for these applications it is necessary that we have understanding and control of the processes used to generate this software. There are many standards, methods and tools available to support the generation of software. Certain standards include requirements for personnel involved in the generation of software e.g. [1], but these are limited to formal qualifications and required experience. This paper however, addresses the human involvement in the software generation process and more particularly the errors introduced by humans into software.

In this paper we discuss how faults are introduced into the software generation process. We have defined the scope of the software generation process for the purposes of this paper as the stages of development between receiving a software requirements specification through to and including coding, i.e. essentially the design activities. We consider the involvement of people with the process and the type of thought processes used by the persons involved.

In particular we consider a model of the fault generation process. This model was developed as part of the Fault Analysis of the Software Generation Process (FASGEP) project which is being carried out within the DTI/IED Safety Critical Systems Programme.

2. Classical Theory and how it relates to the FASGEP Model

Reason [2] indicates that human error is intimately related to the concept of "intent". Error is only a meaningful term when applied to intended (planned) actions that fail to achieve the desired goal without the intervention of some chance or unforeseeable agency. Thus, non-intentional, involuntary and spontaneous actions are not errors. Reason identifies two basic types of error: slips and mistakes.

Slips are where actions do not go according to plan e.g. slips of the tongue, slips of the pen, slips of action. Mistakes occur when the plan itself is inadequate to achieve its objectives. A further category, lapses, are a form of slip, essentially involving a failure of memory which is not necessarily revealed in actual behaviour and may only be apparent to the individual concerned.

Reason's error types can be related to the stages in the cognitive process at which they occur. For the cognitive stages of planning, storage and execution the primary error types are mistakes, lapses and slips respectively.

A further relationship exists between these error types and Rasmussen's [3] classic model of skill, rule and knowledge-based behaviour. Slips and lapses tend to occur at the skill-based level, whereas mistakes occur at the rule-based and knowledge-based levels. Rule-based mistakes are primarily due to misapplied expertise, where some pre-established plan or problem solution is applied inappropriately. Knowledge-based mistakes generally occur due to a lack of expertise, where no off-the-shelf solution exists and an individual is forced to work out a plan of action from first principles.

Reason goes further and highlights the likely failure modes at each level of behaviour. For example skill-based errors occur due to control-mode failures of both inattention and over-attention. Rule-based errors can arise from the misapplication of rules or the application of incorrect rules. Knowledge-based errors can arise due to selectivity, biased reviewing and a number of other factors.

The FASGEP project considers each of the three levels of human behaviour and attempts to define the most significant attributes of the software development process (and metrics for measuring each attribute quantitatively) which influence performance at the appropriate level and therefore lead to errors and the introduction of faults in the software product. These faults may or may not be recovered by review processes within the development life-cycle [4].

It is worth noting, here, that this categorisation and application of attributes relates primarily to Reason's behavioural and contextual levels of human error classification (i.e. it indicates the factors likely to lead to errors of a particular type (slips, lapses or mistakes)), but does not address the conceptual level of error classification, which is concerned with cognitive mechanisms involved in error production.

Reason himself distinguishes between error type and error form. Error forms are recurrent varieties of human fallibility (what others have called psychological error mechanisms) that appear in all kinds of cognitive activity, irrespective of error type.

Thus, the FASGEP approach cannot and does not identify attributes of the software development and review processes which influence human performance in psychological terms (e.g. specific cognitive or socio-psychological factors leading to an increased tendency for focusing and mindset, fixation, overconfidence or other problems giving rise to judgmental and decision errors.) The attributes and associated metrics do, however, identify significant factors in the process which influence the introduction of faults in the software, due to human error. These attributes represent an assimilation of current thinking in recent attempts to obtain measures of fault introduction in software, and are the best judgements of the members of the FASGEP project consortium.

177

3. Fault Introduction Model

The FASGEP analysis model consists of two parts, an inner model to determine the probability of the number of faults generated or removed by a particular atomic process, and an outer model which propagates the results from the inner models as the project progresses through the software development life cycle and results in a probability distribution for the estimated number of faults remaining at the end of the development life cycle [5].

The FASGEP project has identified two classes of graphical probability models (GPM) for the inner (or fault introduction) model:

1) The development process which introduces faults
2) The review process which identifies faults which in the semantics of the FASGEP project is responsible for fault removal.

The purpose of the development fault introduction models in each atomic process is to calculate the probability of faults being introduced during that atomic process. This probability is in the form of a distribution over fault numbers and known as fault propensity. At the highest level the total fault propensity is calculated from a convolution of the fault propensity for Symptomatic faults with the fault propensity for Random faults, but the GPMs are also used to perform completely general Bayesian updating calculations. The GPMs enable fault propensity to be calculated from a fusion of the prior probability distribution for each net, and evidence data collected from metrics of the generic atomic process attributes which are observed for each atomic process.

It is the development fault introduction model which assess the possibilities for human errors. A GPM allows a collection of factors, and the probabilistic relations between them, to be modelled. The structure of the networks are shown below, on which a GPM is based describes which variables influence each other, and conditional probability matrices are used to weight the various influences.

Symptomatic faults are those faults which have been identified to be caused by a defect in the process, rather than the people carrying out the task. It may be that different atomic processes by their nature will exhibit different failure characteristics; the project has been careful to maintain a generic model of software development, without specifying a particular process or life cycle. The factors influencing symptomatic faults were identified in the process modelling activity as: Goodness of (process) interfaces which includes undefined processes, Project Management Quality, Quality of Input Product and Goodness of Method.

Random Faults in the context of FASGEP have been identified as being associated with faults introduced by human behaviour.

3.1 Skill-Based Errors

Skill-based errors are identified by problems due to inattention or over attention to the specific task. These are typically identified by slips, omissions or repetitions in the product being produced. The causal net for skill-based errors uses attributes of ability, motivation and environment quality (see figure 1).

The measurement of a person's intrinsic ability to carry to a specific task is difficult to determine. One method is the use of psychometric tests e.g. [6]; however, this method was not considered appropriate for the initial case studies in the FASGEP project and a subjective estimate of ability was used.

Job satisfaction, morale and workload match, contribute to self motivation and are closely inter-linked and it would be difficult to recognise a situation in which all three factors are not equally important. All three factors can vary: with time; between individuals; and as a team. It is recognised that high levels of job satisfaction and morale can result in high levels of self motivation, but the relationship is very complex and can be influenced by other personal factors.

High levels of job satisfaction are generally only achieved when the basic needs of security and salary are satisfied. There are occupations where salaries are low but job satisfaction is high, but the norm is the reverse. Low salaries and job insecurity can be a source of stress factors, which have a detrimental effect on overall job satisfaction. High salaries and job security alone may not be enough to produce high job satisfaction. Other more complex emotional factors are required for high levels of job satisfaction.

Job satisfaction and team interaction are significant factors in determining morale in the working environment. Recognition of skills, and of being a valued member of a team, also contributes to high morale. Morale can be adversely affected by lack of recognition, inability to communicate errors freely, isolation and general lack of cohesion within the peer group. If these negative factors are present in other team members, team morale and motivation would be expected to be low.

Working Environment Quality is a complex problem to measure. A separate causal net was developed for this quality. The working environment quality is influenced by many attributes: by the Workstation Quality, Effectiveness of Communication, Working Environment Satisfaction and by the presence (or absence) of individual control over the working environment.

The Workstation Quality of the individual team member's own working area (e.g. their allocated desk or wherever they spend most of their time) has several attributes: the comfort of the individual is felt to be the overriding factor.

The comfort factor assesses the quality of the ambient environment. In this assessment it was felt that the following attributes were considered important: lighting, heating, noise and ventilation.

The European Commission [7] have set down statutory requirements for persons working on computer terminals. These requirements are reflected in the quality of facilities factor.

The team members' satisfaction with their working environment entails both the level of distraction to which they were subjected and the quality of the services provided. Distractions are known to have a severe detrimental effect on the ability of an individual to perform a given task accurately. The level of distraction is found from attributes associated with the number of neighbours not contributing to the task in hand and the number of neighbours in total. Noisy items, especially intermittently noisy items, contribute to the level of distraction In determining the quality of the services provided to the team members, the availability of services is obviously the overriding factor. The secondary factor is considered to be the quality of the service provided and finally the proximity of the said services.

3.2 Rule-Based Errors

Rule-based errors occur because of the working method and procedures imposed. Poor working methods and procedures can give rise to disagreement, disillusionment or even resentment from the workforce. This may be because the rules are: inelegant, inadvisable, too strong, or too general. It could be argued that a professional engineer would not make this type of error but, even so, there may be problems with interpretation, or implementation, that would lead to this type of error being produced. (The causal net is shown in figure 2)

Task appropriateness and the quality of project management have been considered to be the main attributes for rule based errors.

A task can be regarded as appropriate when an individual or a team have the necessary skills to complete the task satisfactorily and the task has been adequately defined. Inappropriate tasks may be recognised by the level of errors that appear in the task life cycle. Experience-level and task-experience matching are seen as the most significant factors in task appropriateness.

Four factors have been highlighted as having significant influence on project management quality. Leadership is a contributor to good quality management since

180

it can demonstrate commitment to a project from the top of the organisation. Team quality (continuity, cohesion and team size) and effective communications are also important.

Team attributes are those characteristics that produce co-operation, camaraderie and good internal communications which tend to lead to a "quality" team output. A good team has a capability level greater than that achievable by the efforts of the team members acting as individuals. Team cohesion and continuity have a more significant effect than team size.

3.3 Knowledge Based Errors.

Rasmussen [8] has indicated that experience is the most important of the attributes in determining the number of knowledge-based errors. A good team has been shown [9] to result in a reduction in the number of errors introduced. Shared experience has therefore been considered to have the highest weight in assigning the probabilities in this distribution (see Figure 3).

Task comprehension has been determined from the questions on task familiarity and an adequate task description. The experience attributes are obtained from the responses to multiple questions on experience. The shared experience factor determines the effect of the team working together to pool knowledge in solving problems. An individual working as part of a cohesive team can use the experience of others, and conversely the lack of cohesion within a team prevents experience being shared [9] [10]. Cohesion in this context has been determined from social interaction, team dedication and the number of new members joining the team .

3.4 Communication

Communication is an important human factor in the generation of software errors, and its effects can be seen in both random and symptomatic types of faults. In the FASGEP project we have developed a causal net for the effectiveness of communications which is used in both random and symptomatic nets.

The effectiveness of the communication attribute varies according to the complexity rating of the hierarchy. For simple hierarchical structures, informal communications are considered more appropriate than formal ones since the team is likely to be small with little problem in discussing issues with their peers and superiors. Complex hierarchical structures however, require that communications be more formal to ensure that the correct people are kept informed of all relevant (and only relevant) developments.

To assess the complexity of the reporting hierarchy, data is captured detailing the number of reporting levels and the number of sites involved in the development. During the development of the Causal Network, it was realised that a further measure of complexity is to be gained from capturing the number of lateral paths in the reporting structure.

In determining the Effectiveness of Formal Communication, it was decided that the flexibility of the communication within the team was the best indicator of effective communication, the type and frequency of team meetings were considered to be of similar importance, and the quality of feedback to team members was considered to be a relatively minor indication of effective formal communication.

The main indicator of effective informal communications was decided to be the quality and quantity of verbal communication in preference to non-verbal communication.

4. Data Collection

It has been shown many times that the collection and storage of data is a vital aspect of any unification framework. Poor data-collection techniques and requirement definitions have been the causes of the limited success and acceptance of several other projects aimed at developing predictive models for software development.

Although the general requirements for a successful data-collection exercise include automating as much of the process as possible using formal measures, it was realised early in the FASGEP project that human factors were a large contributor to the introduction of faults in software. These are expected to include factors such as the individual's job satisfaction and morale which can realistically only be determined by direct questioning of the individual. This requires the use of a questionnaire approach.

Additionally, since FASGEP requires the collection of data from several sources, each of which uses different development processes and fault collection techniques, it was decided that to collect such data by automatic means was at this stage impractical and would impose severe restrictions on the data available to the project. Thus, the questionnaire approach was deemed the most appropriate technique at this stage of development of the predictive model.

Questionnaires have several inherent weaknesses that the FASGEP project recognises and has attempted to minimise by careful design:

(i) Question wording directly affects the validity and reliability of a questionnaire

[11].

(ii) The format of the questions is also important: should they be open or closed. Closed questions can force inappropriate response, but are easier to capture and check.

(iii) Since questionnaires rely solely on the interpretation and feelings of the respondent the answers may be biased and may exhibit some degree of subjectivity.

(iv) Respondents are sensitive to the context in which the question is asked, as well as the particular words used to ask it. As a result, the meaning of almost any question can be altered by a preceding question [12].

As already stated, it was an obvious requirement that the project be able to collect data on the feelings and opinions of the individuals involved in the software development. In cases such as these, the approach of using a questionnaire actually becomes beneficial with respect to the normal problem areas of subjectivity and bias. It is these aspects of human emotions that we are interested in capturing and using as evidence of the motivation and satisfaction of the individual.

5. Case Studies

The aim of the FASGEP case studies is "to provide vehicles for testing the FASGEP Model and Method as they evolve throughout the project". That is, they will be used for calibration [1] and verification [2]. Three types of case study projects are being used by the FASGEP project.

(i) Past Projects

Projects that have been completed and delivered to the customer. Although the fault data can be considered to be complete, the collection of the human factors data is difficult since the development team is likely to have spilt up. Also much of the other data required by FASGEP is unlikely to have been captured during the development.

(ii) Ongoing Projects

These are projects in which the development is already some way to completion. Although some data may have been lost, these are useful to observe the effect of introducing new collection requirements on the team both in terms of attitude and resource requirements.

---

[1] Calibration: the process by which the parameters of the model are adjusted to provide a result that fits observed data accurately.
[2] Verification: the process of ensuring that the model works correctly

(iii) New Projects	New projects are those for which the FASGEP data collection scheme is implemented from the very start of the development. Thus all data is available. These allow the identification of the changes in the team throughout the development and how these changes can affect the integrity of the final software.

6. Results

At the time of writing (August 93) none of the case studies of the type described above have completed the data-collection exercise. However some initial analysis of the fault reports in two case studies were available; these are shown below:

Test Case A

Application: Aerospace - Fuel Control System
Language: Proprietary (Assembler)
Lifecycle: Structured - with incremental delivery
Sample size: 45 errors - of which 11 Symptomatic; 27 Random; 1 unclassified; 7 not faults

Test Case B

Application: Aerospace - Integrity Monitoring System
Language: Ada
Lifecycle: Evolutionary
Sample size: 28 errors - of which 12 Symptomatic; 14 Random; 2 not faults

The random faults were further analysed to estimate the cognitive level of the error:
Skill based = 2; Rule based = 2; Knowledge based = 9; there was one fault where there was insufficient information to categorise.

It is hoped that by the time the paper is presented, the results from several case studies will be completed and these will be reported.

7. Conclusions

To improve the software generation process, it is clear that account must be taken of the humans involved in the process. This is in addition to the tools, methods and procedures used. The FASGEP method provides a means of estimating the number of detectable faults in a product caused by human error using the objective and subjective information available.

A potential use for this method could be for testing scenarios before changes are made to the development teams and working environments are made.
Further work is now required in validating the method using case studies.

8. Acknowledgements

The authors gratefully acknowledge the partial funding provided by DTI under the Safety Critical Systems programme reference IED/1/9004.

The FASGEP consortium consists of Lloyd's Register of Shipping, Lucas Electronics, Lucas Engineering & Systems, Nuclear Electric, The Open University and Rolls-Royce.

The authors would also like to thank Sarah Grey, Dr Lesley Winsborrow, Steve Connolly, Martin Cottam, Reg Parker, David Nicoll, Jon Speer, Nick Bird, Dr Eddie Williams and Dr Mike Falla for their considerable contribution to the project

References

[1]   Defence Standard 00-55 interim issue 1 April 1991
[2]   Reason J. Human Error. Cambridge University Press Cambridge, 1990
[3]   Rasmussen J. Skills, rules, knowledge: signals, signs and symbols and other distinctions in human performance models. IEEE Trans Systems, Man and Cybernetics SMC-13 (1983), 257-267
[4]   May J, Hall P, Zhu H, Cockram TJ, Bird N, Fault Prediction for the Software Development Process. Proceedings of IMA conference on the Mathematics of Dependable Systems, Egham, Surrey September 1993
[5]   Hall P et al., Integrity Prediction during Software Development Proceeding of Safecomp'92 conference, Zurich October 1992
[6]   Aiken LR, Psychological Testing and Assessment, Allyn & Bacon 1988
[7]   European directive 90/270/EEC 29 May 1990
[8]   Rasmussen J. Information Processing and Human Machine Interaction North Holland, 1986
[9]   DeMarco T, Lister T. Peopleware. Productive Projects & Teams Dorset House 1987
[10]   Basili V, Reiter R. An investigation of human factors in software development Computer Dec 1979: 21-38
[11]   Ed. J Richardson. Usability Evaluation, RACE project deliverable (ISSUE programme) Nov 1992
[12]   Converse JM, Presser S. Survey Questions - Handcrafting the standardised questionnaire Sage Publications, 1986

Cognitive and Organisational Aspects of Design

Martin Loomes, Donald Ridley and Diana Kornbrot
The University of Hertfordshire, Hatfield, UK

Abstract

Research into the software design process currently centres on a particular model of the software design which is based on a number of assumptions that are rarely tested, and have little theoretical grounding. This paper attempts to highlight some of these assumptions and to suggest ways in which they might be limiting current research activity. It identifies the life-cycle as the core of the existing paradigm, and introduces an alternative model that may be more fruitful for the discussion of cognitive and organisational aspects of the design process.

Background

The risk of failure has always been recognised as an inherent part of design, particularly when new technologies are being applied to novel problem domains [1]. This failure may manifest itself as visually unattractive buildings that ruin a city centre, stylish chairs that are impossible to sit in, or railway systems that grind to a halt whenever leaves fall onto the track. Systems that involve software are also designed, of course, so it should come as no surprise that they too are liable to fail. What is surprising, however, is the myth that seems to be developing around software design that we can find ways of avoiding these failures. One possible cause of this is the belief that because software is not a "real physical entity" there is no excuse for systems failing due to software errors. Whilst we can accept that metal fatigue may cause an aircraft to crash, there seems to be the suggestion that failure due to faulty software is more easily avoided. The assumption seems to be that we can construct software that is "correct", if we find and apply the "correct" methods for the task. We seem reluctant to recognise that the cause of system failure is not metal fatigue or software errors *per se*, but the designer's decision to use metal or code that is prone to fail: it is the designer that causes the failure, not the materials. Rather than seeing failure and errors as things that exist, but can be avoided with the right methodology, we can view them as things that the designer brings about, and ask what behaviour causes this. If we understood better why designers make mistakes we might be able to suggest ways they can adjust their behaviour to minimise errors, or contain their impact on the process as a whole.

It has been suggested that primitive societies are reluctant to allow designers to experiment with new approaches, as they live the sort of hand to mouth existence where any unforeseen cost might be catastrophic. As a result

> "... primitive societies are very conservative. Tribal customs prescribe
> exactly how everything shall be done, on pain of God's displeasure. An
> inventor is likely to be liquidated as a dangerous deviationalist." [2]

Although we cannot describe our western civilisation as primitive, it might be suggested that we are now so dependent on our designed systems, and they are so complex and all-embracing, that we are demonstrating similar attitudes. The consequences of failure are catastrophic not because of the poverty of the environment in which they occur, but because of the richness of the role the systems play in our lives. This poses very real problems for software designers: the systems they are being called upon to design are often very complex, and need innovative approaches, but the environment in which they are working is reluctant to allow them the freedom to take the necessary risks. System failures that do occur usually serve to reinforce the intuition that more control, rather than less, is required. Moreover, the usual prejudice that control implies rules, and centralisation, and cannot be an emergent property of the design process itself, is usually applied, so that responsibility is removed from the designers to higher agencies such as departments, companies, authorities and governments. When failures occur, however, as they surely will, they are usually seen as local to a project, rather than a product of the culture within which the design is taking place. The fixes that are applied are local, *ad hoc*, solutions to an immediate problem, and rarely percolate through to inform the wider society in which design is occurring. Software engineers are constantly being told that they should not simply fix bugs at the code level, but should address the higher levels of design too: this does not usually include addressing the problems with the methodology and tools which led to the problem! They may debug the artefact, but not the process.

Responsible, Self-conscious Design

Alexander has suggested that a crucial component of modern design is the realisation that the designer must be self-conscious, and be prepared to discuss and change the process of design as well as the artefacts produced [3]. This is because modern design is, in general, not carried out to solve problems of immediate concern to the designer, but on behalf on others. Moreover, it is usually carried out by teams of designers rather than by individuals. Thus communication of ideas is crucial to the process. Primitive man, however, can proceed "unconscious of the fact that among his faculties there is one which allows him to refashion nature according to his desires" [4]. He can apply the tried and trusted methods of his ancestors to the everyday problems he encounters, and, should they fail him, he can deviate slightly and gain immediate feedback as to whether or not the deviation is helping or hindering the process of solving the problem. The computer hobbyist, of course, can use very similar approaches to "hack" solutions to problems on a home computer. Many computer scientists would have it that this is a "wrong" approach to software design: in fact, of course, it is simply a technique that does not scale up beyond the single-designer, immediate-feedback, paradigm. One of the consequences of accepting that software design must be self-conscious is that the designer must accept the loss of innocence that this entails [5]. Alexander cites two reactions from designers who are unwilling to accept this loss of innocence, particularly when faced with complex tasks that they feel unable to handle: the refuge in genius and the refuge in style.

Refuge in genius might be viewed as acceptance of the idea that there is a Muse of design who provides the inspiration behind the process: this is an escape from the loss of innocence because clearly the designer cannot be held to blame if the Muse gets it wrong! Although few people would admit to holding this view, there are certainly many who see intuition or inspiration as central to the design process, and would argue that attempts to be scientific in design are dangerous because they cause designers to inhibit their intuitions (to close their minds to the Muse).

The refuge in style identified by Alexander is the attempt to carry out design within a particular style or school. The designer who adopts an "art nouveau" or "Georgian" style, for example, and creates a monstrosity, can seek refuge amongst like-minded designers, and argue that it was the style that failed, not the designer. This is a far more worrying trend amongst software designers! Top-down design, stepwise refinement, formal methods, object oriented design, and many more, can all be seen as styles that designers might work within and try to escape the loss of innocence. If they choose to work within a style, of course, they are not escaping at all, but should be accepting responsibility for the choice of style as part of the design process. Unfortunately, education often legitimises this escape route by introducing "methodologies" in such a way as to suggest that they are useful without careful consideration of the task for which they are being selected.

If we put together the apparent reluctance of the software design community to take risks when developing systems, and the use of refuge in style to escape the loss of innocence, we have a potentially disastrous situation. Industry wants designers to use "tried and tested" methods and designers are often only too happy to abrogate responsibility for the choice of method and work within a house style. Moreover, to make the designer more "efficient" the style is often buried deep in the workings of a CASE tool, so that the designer does not even have to be aware of the processes involved. Unfortunately, the tried and tested methods we have are only useful for technologies and problem domains which are well understood. Although there are still many design tasks to be carried out in these areas, the real challenges are in areas where there are no such aids.

We are currently deeply wedded to a culture in which the true nature of design is buried beneath the quest for the methods we need to avoid failure. Companies that have recognised the severe limitations of such methods for most real problems dare not say so too explicitly, for the software house that advertises with the slogan
"We don't use any particular method or tool"
will certainly find itself upstaged by the company who uses OOD and the most trendy CASE tools. Individual designers who constantly question a house style may well find themselves promoted to "design consultant" positions, where they can be isolated as deviationalists if necessary, or put in charge of "special projects", but will rarely be made project managers for mainstream developments. We would argue that while the quest for such methods dominates research, and their application dominates practice, we will never significantly improve the ways in which we design software systems. We may well "polish the shiny bits", making aspects of design we choose to see rather more acceptable, but we will not tackle the messy problems that lie beneath the surface.

Technocentrism

This sort of global criticism is nihilistic, of course, unless we also take the bold step of putting forward some alternative ways of proceeding. We have denied ourselves the easy route of simply proposing another methodology, or encapsulating existing ones into more powerful tools, and so we need something rather more radical. The avenue we wish to explore is to see what insights can be gained by rejecting the notion of a development life-cycle as the dominant feature of the paradigm in which we work. Rejecting the life-cycle as the foundation upon which we want to build effectively eliminates most existing methodologies at a stroke, for most of these are prescribed routes and notations for navigating around a life-cycle of some form.

Moreover, we suggest that concentrating on code as the end-product of software design may be placing emphasis in the wrong place, allowing the technical aspects of the problem to determine the paradigm within which design will take place. Papert uses the term "technocentrism" to capture to the idea that we refer all questions to technology [6]. Adopting a life-cycle model epitomises the way in which technocentrism is the norm in software design. It suggests that the life-cycle is an emergent property of some naturally occurring phenomena. The quest for more accurate models appears to be interpreted by many as the search for a better understanding of this essential property. In fact, as Papert notes, technology does not have to be considered in this way: we can, if we choose, recognise that we control the technology. Rather than asking questions such as "how does inheritance work in

189

object oriented design" we can ask "how do we want to use the notion of inheritance when we design systems". If we want to consider not only the technical problems, but also wider issues such as how software designers might be made more effective, how users should be integrated into the design task, what organisational structures should we put in place that will have a beneficial effect on the task, and what cognitive processes are involved in design, we must adopt less technocentric views of the process. If we do not adopt such views we sacrifice control over how the questions are posed, and we often have to accept that some questions cannot be expressed at all. We should note that the term "safety-critical systems" itself invites a technocentric view, for it is often taken as the classification of systems that are safety-critical, rather than systems which we are making safety-critical by the way we intend using them.

It is important to note that we are not advocating replacing one model of the process with another: rather that we should be encouraging researchers to choose new ways of viewing the design process to supplement and stimulate the current models.

The Theory-Building View

One model that deserves more detailed analysis is the Theory Building Model. Here we do not refer questions to the technical product of the design task, but rather we anchor discussion in the production of theories, which have the desired system as a model. Exactly what comprises a "theory" in this context, of course, is an essential part of the exploration, and not something that can be given a glib answer here. We should note, however, that blind acceptance of the dated view that theories are implicit in nature, and the scientist's task is to discover them, will lead us straight back to technocentrism: rather we will assume that theories are the attempts of mankind to impose some order onto phenomena. They are thus designed artefacts in their own right.

This idea is not new. It has been proposed by at least two other sources. Burstall and Goguen, for example, suggested that we can put theories together to make specifications, where the theories are captured in a suitable logical system [7]. Naur, on the other hand, suggested that programming can be viewed as theory building, where the theories are in the minds of the designers [8]. It is interesting to note, however, that these two ideas are rarely seen cited in the same literature. The former in now seen as part of "formal methods" (although it originally appeared in the AI literature), whilst the latter is more closely associated with the "softer" aspects of system design. The current paradigm within which Software Engineering research is being carried out makes it very difficult to reconcile these two views. Formal methods and intuition are seen as opposite poles in some implicit construct. We would argue that they are both essential tools to be used by a software designer, but that we need to explore how they can be reconciled.

Cognitive and Organisational Processes

Consideration of the theory-building view raises some very interesting, and important, questions. In this section we will outline some of these, and suggest possible avenues of research that might be used in the quest for answers.

If primary outcomes of the software design process are theories we should consider the relationship between software design and science, which also has theories as a major goal. A "scientific approach" to software design has been advocated in many places before, but usually the benefits claimed for such an approach seem undeliverable. The approach has been presented as a way of ensuring "correctness". In fact modern science has given up all claims to be discovering correct theories, recognising that all theories, if scientific, are potentially refutable. The scientific approaches suggested for software design certainly can deliver "correctness", when the term is given a severely limited meaning that divorces software systems from the real world, but most designers recognise that this avoids the really interesting issues and areas where failures usually occur. Unfortunately, this has caused many designers to reject any discussions of science as relevant to the design process, as they assume that similar simplifying assumptions will be made. We would argue that a genuine attempt to relate scientific practice to software engineering practice will provide some powerful insights that might lead to real improvements in the design process. This can be achieved by examining, and rationalising, the ways in which scientists work in practice: an endeavour which has been carried out by philosophers of science for many centuries. Viewing design projects in the context of Kuhnian paradigms [9], for example, provides a way of discussing the interaction between projects, something that cannot be achieved in most models of the software design process. Extending Kuhn's ideas to those of Lakatos [10] allows us to recognise hard cores of the design process that serve to form paradigms: areas that will rarely be challenged by those working in specific problem domains.

This leads us to question the source of such paradigms and hard cores. Some of these are explicit, being axiomatic in the problem domain or in the methodology that the designers have been told to adopt. Many more are implicit, however, being hidden in the tools used for the job, the education of the designer, or the culture the designer in working within. Examining such issues is a difficult task, and it is easy to see why technocentrism is often the preferred route, as we only have to ask "hard" technical questions, rather than these "soft" questions which we don't expect to have satisfactory answers. It is also easy to see why many institutions and companies would prefer technocentric answers: it is much easier to accept that you need to buy a new tool or impose a new methodology, than to risk asking why your company organisation is such that the designers are making a mess of so many projects. Never-the-less, if we want to do more than "polish the shiny bits" we believe that such difficult questions need asking. There are approaches to the systematic investigation of these sort of questions that may be fruitful, typically those drawn from occupational and organisational psychology. The cultural audit [11], for example, may provide a way of carrying out the type of holistic study of a design team that is necessary to start exploring the human causes of error. No doubt it will need tailoring in order to extract the right sort of information, but early studies suggest this might be feasible [12].

Once we have taken the bold step of realising that software design is a human process, rather than a purely technical one, we can also start to ask questions about how individual designers interact and make decisions. The life-cycle model assumes that there is a single specification and a single design, whereas in reality we must assume that each designer will place different interpretations on these, even if document control is sufficiently draconian to ensure only one written form is permitted. Current approaches attempt to impose highly structured notations and diagrammatic forms on the process, so that we limit what can be said. The implicit assumption seems to be that this will reduce the problems of different theories in the heads of individual designers. There appears to have been little or no research into such claims, and little research into the extent to which the imposition of notations reduces the ability of designers to communicate the theory that is being constructed effectively. There has also been very little research into how designers make judgements in the course of a design project. Clearly a good understanding of this is essential if we want to improve the process by suggesting ways in which we might help them to avoid judgements that lead to system failure. Achieving this is a nontrivial task, but to ignore it because it is difficult is to avoid the crux of the problem.

It is important to recognise the interaction between the cognitive and organisational aspects of the problem. We hope that by integrating both aspects in one research programme we can achieve a constructive tension between the two, making hypotheses in one area that we can test in the other.

Conclusions

In this short paper we have attempted to raise the question of the wisdom of centring all research into software system design on a single paradigm, the life-cycle model, which seems to have arisen with little supporting empirical or theoretical evidence. We have also introduced another model that we believe opens up the possibility of exploring a number of research questions which are currently not sufficiently represented in Software Engineering research. In particular, by taking a less technocentric view we are able to discuss the cognitive and organisational aspects of design in a way that makes them central to the human activity of design, rather than relegating them to peripheral issues that hang from the life-cycle of the artefact.

References

[1] Petroski H. To engineer is human, Macmillan, 1982

[2] de Camp L S. Ancient engineers, Tandem, 1977

[3] Alexander C. Notes on the synthesis of form, Harvard University Press, 1964

[4] Ortega y Gasset J. Thoughts on technology. In: Mitcham C and Mackey R (ed) Philosophy and technology, The Free Press, New York, pp 290-316, 1972

[5] Loomes M. Selfconscious or unselfconscious software design?, Journal of Information Technology, 5(1): 33-36, March 1990

[6] Papert S. A critique of technocentrism in thinking about the school of the future, E&L Memo No. 2, Massachusetts Institute of Technology Media Laboratory, September, 1990

[7] Burstall R M and Goguen J A. Putting theories together to make specifications, In: Proceedings of the Fifth International Joint Conference on Artificial Intelligence, 1977

[8] Naur P. Programming as theory building, Microprocessors and Microprocessing, 15, pp 253-261, 1985

[9] Kuhn T S. The structure of scientific revolutions (second enlarged edition), University of Chicago Press, 1970

[10] Lakatos I. Mathematics, science and epistemology, Philosophical Papers 2, CUP, 1976

[11] Fletcher B. The cultural audit, an individual and oganisational investigation, CPI, 1989

[12] Crimes M. The safety culture audit, MSc Thesis, The University of Hertfordshire, 1992.

Producing Critical Systems —
The Ada 9X Solution

B A Wichmann

National Physical Laboratory,
Teddington, Middlesex, TW11 0LW, UK

Abstract

Using an appropriate programming language for safety-critical systems is important, since it is necessary to demonstrate that the actual system has the required properties. Here we show that the revision of the Ada programming language specifically addresses the concerns of the safety community by providing greater visibility of the actual object code and by allowing the user to specify a subset of the language which is enforced by the compiler. Hence it is believed that Ada 9X will be the most appropriate language for applications involving safety.

1 The Problem

We all know that writing critical systems is difficult and expensive. The problem is not producing a system which appears to function correctly, but in providing adequate justification that the system is suitable for such a critical application. Several parties may need to be assured of the quality of the software, such as the line manager, quality staff, company lawyer, certification body and Government regulatory agency. All their concerns need to be understood and responded to. This is an inevitable consequence of the risks involved which are well covered in a recent Engineering Council publication [4].

In software engineering terms, the issue is one of producing an agreed functional specification for the software component and then demonstrating that the code produced satisfies this specification. Since producing an unambiguous specification is almost impossible, and providing a fool-proof demonstration is a practical impossibility, many engineering compromises must be made and agreed with all the relevant parties.

The Ministry of Defence standard which is based upon a mathematical specification and formal proofs or rigorous arguments for compliance has a lot to recommend it [5]. However, the complete application of this standard is often not feasible and hence other methods must be used, even within the defence context. Whatever approach is taken, the form of the eventual program is vital, since it is an analysis of that which must provide the basis of most of the assurance procedures. This implies that the programming language used for the application is likely to have a key role.

A mature standard in the safety-critical area which takes a different view to formal methods is the internationally-agreed avionics standard DO-178B [8]. Here the emphasis is on design, analysis and test. Although there is little stated about programming languages, it is clear that any analysis will either depend upon the programming language or upon a corresponding analysis of the object code generated by the compiler. Quite rightly, the standard requires isolation from the correctness of the compiler for the most critical software which implies that one must reason from the object code or show the object code is equivalent to the source code.

Standard Language	Security features	Insecurities left
Ada	Run-time checks required. Type-secure across packages. Can recover from check failures.	Access to unset scalar.
Modula-2 (not yet ISO standard)	Type-secure across modules. Limited recovery from failures.	Unset pointers (& scalars).
Pascal	Strongly typed.	Run-time checks optional. Unset pointers (& scalars).
C	(Additional tools: make and lint).	150 undefined 'features' Run-time checking typically not done.
Fortran 77	Type checking. No pointers.	Default declarations. No checking across routines.

Figure 1: Language Summary

2 The language context

To reason about a program requires that the structures used within the program are well-defined and properly implemented. Unfortunately, almost all existing programming languages standardized by ISO are not defined in a mathematically precise form. Hence substantial care must be taken to ensure that the features of the language used are well-defined and that the language definition accords with the intention of the programmer.

Since we are all fallible, even programmers, languages which provide checking for some forms of error are an advantage, especially if the design can allow the software to request the system to return to a safe state. Ada is the only widely used language suitable for critical applications which requires such checking. Many critical applications do not exploit this checking but demonstrate (perhaps by mathematical proof) that the checks could not fail. This form of static checking is very expensive in staff time to undertake, and hence is not practical for less critical applications. For a summary of the securities and insecurities of standard languages, see Figure 1.

A comparison between programming languages undertaken some time ago showed that a subset of Ada was a good choice [3]. The subsequent maturity of Ada compilers, the better understanding of the language and the provision of special tools for Ada, makes the language the first choice for many applications. In contrast, the 'C' language is deprecated in the IEC draft standard on safety software [7] since it is very difficult to demonstrate that 'C' code is restricted to the subset which is well-defined.

The validation problem is illustrated by Figure 2 in which the relationship between the crucial subsets of source programs is shown. The ideal would be if all programs which compiled had a well-defined meaning. In the figure, this ideal would be if the area in grey did not exist — grey is the danger area for critical systems. Ada comes quite near to the ideal, whilst C makes no attempt to reach this position. Even if a program has no well-defined meaning according to the semantics of the language, it may function correctly due to properties of the implementation which are not guaranteed. Relying upon such properties is unwise, since even a compiler upgrade could produce different results. Unfortunately, showing that a program does meet all the requirements of a standard is tricky. The approach taken in [8] would be to rely upon high quality testing,

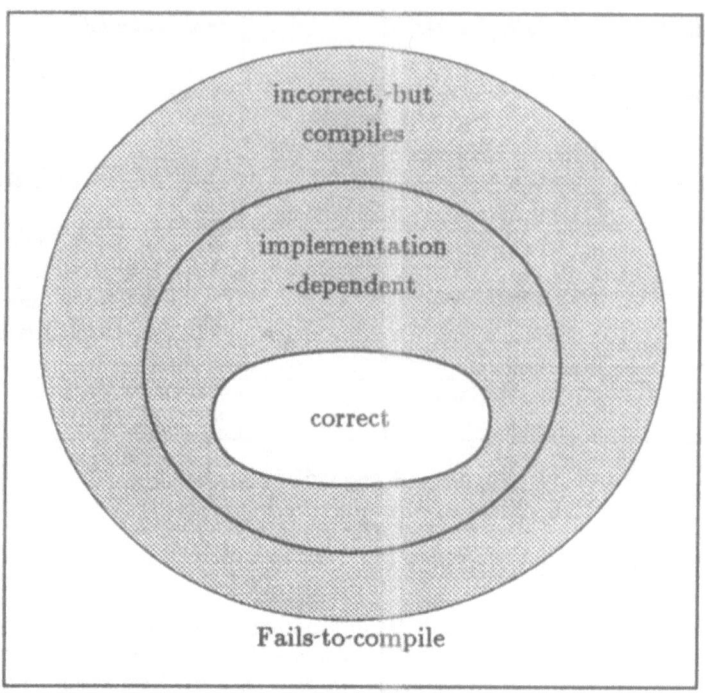

Figure 2: Language subsets

which is again hard to do in a manner which is totally convincing.

Nevertheless, problems have been noted with the Ada language definition some of which can make it difficult to validate code [9]. These issues are carefully avoided by suitably chosen subsets of Ada, such as that provided by SPARK [2]. The same subset could be used for Ada 83 and Ada 9X.

An important aspect of Ada is that it provides support for the provision of secure abstractions. For instance, an Ada package can provide a type `Message` with all the operations on this type in a manner that cannot be subverted by users of the abstraction. This is because of the strong typing in the language, and because it is possible to control the type completely. In contrast, 'C' does not provide strong typing, and Modula-2 does not provide a method of initializing variables of an opaque type. In Ada 9X, the abstraction mechanism is extended to include further support for the object oriented programming paradigm.

Until now, language standards have not addressed the problem of the validation of the object code generated by a compiler. This is troublesome unless information is provided linking the source code to the object code. Such checking will be required for many years to come while there is a possibility of errors being introduced by a compiler. The alternative of having 'trusted compilers' does not yet seem viable for the major languages.

196

3 The Ada 9X solution

The ISO standardization process for Ada requires that a periodic review is undertaken of the standard. In 1988 it was agreed that a revised language should be produced, both correcting known minor problems in the language and including new features demanded by application concerns and software engineering issues [6].

An overview of the facilities in Ada 9X is given by John Barnes in [1]. Before introducing the additions for critical applications, we describe the main features new to Ada 9X.

3.1 The 9X enhancements

The main user needs identified as part of the revision process in which several hundred comments were received were as follows:

Interfacing. Ada 83 was designed to produce both embedded systems (built upon bare boards) and conventional systems working on top of proprietary operating systems. However, since 1983, systems have become increasingly sophisticated by providing Graphical User Interfaces. Programs must be able to interface to such systems in a convenient fashion to aid system development and enhancement. Ada 83 was also defective in specifying characters as being exactly 7 bits, thus inhibiting the use of extended character sets.

Programming by Extension. Systems must be able to grow without major re-design. The Ada 83 foundations of packages and generics is excellent, but experience has should difficulty in extending systems without at least recompilation, as we shall see below. Although Ada 83 does support Object Oriented Design, the conversion of the design to code could be improved to permit extensions.

Increasing flexibility with program libraries. The Ada program library mechanism provides type-safety across separate compilation in a natural manner. However, for larger systems, it is necessary to provide something more than a flat library structure. The existing structure cannot directly support a system involving more than one target machine.

More efficient tasking. The Ada 83 tasking mechanism based upon the rendezvous is very elegant but is inefficient when compared with hand-producing tasking kernels in assembler. An example of this is the requirement to control concurrent access to data without incurring rescheduling. Indeed, even within the Ada 83 model, there is a need to be able to control scheduling to take into account theoretical advances, such as rate-monotonic scheduling.

In each of the above areas, we consider how Ada 9X has been designed to satisfy the requirements while maintaining upward-compatiblity with Ada 83.

3.1.1 Interfacing

The basic mechanism of the ability to call routines external to the Ada program is present in Ada 83. However, there are cases in which Ada 83 is inadequate or not portable. A major interface that is a separate Ada standard is that for access to Posix. The Posix Ada binding could be a vital interface for application programs. Also, within

197

the Ada 9X standard itself, packages are defined for basic numerical calculations, so that portability is increased by not requiring access beyond the Ada library.

For facilities such as Windows environments, a call-back facility is needed in which a procedure parameter is passed to the interface. Since no procedure values were available in Ada 83, this could only be undertaken using detailed knowledge of an implementation. In Ada 9X, access types are extended to include access to subprograms so that appropriate values can be constructed for passing to such interfaces in a type-secure fashion. The calling convention for such procedures can be specified.

At the lower level, interfacing to hardware is facilitated by pragmas which can control the use of a variable or a type. For instance, with memory mapped input-output, an Ada variable may represent a device control register. Access to this register must be exactly as written in the source program, rather than in a manner determined by the code optimizer.

3.1.2 Programming by Extension

A common problem is to add a small functionality to a large system. If such functionality is merely adding to an existing procedure, then the separation of package specification and body provides all that might be required. In particular, the existing body can be replaced by the extended version without recompiling the entire system.

In practice, extensions are rarely that simple. In particular, it is a frequent requirement to extend the fields of a record. For Ada, this is a major problem, since the language is designed for high performance execution which implies that indirect access to data (to allow for any additional fields) should not be required.

The Ada 9X solution to this problem is to provide a new class of types called **tagged** types, which can provide the needed flexibility.

To illustrate this, consider an application in graphics:

```
type Point is -- a conventional type which cannot be extende
   record
      X, Y: Float;
   end record;

type Line is tagged -- an extensible type
   record
      Start, Finish: Point;
   end record;

type Fancy_Line is new Line with
   record
      Width: Float;
      Miter_Value: Float;
      Style: Dash_Encoding;
   end record;
```

Since we have indicated that type Line is tagged, any package which uses this type will also function with Fancy_Line values by simply ignoring the additional fields. This is important, since we do not wish even to recompile code working with Line merely because we have chosen to provide the ability to handle Fancy_Line.

As a further example of type extensions, consider a system which handles alarms in different ways according to some classification of the alarms. The traditional method would be to introduce the classification at the highest level, so that the introduction of a new class causes the entire system to be at least recompiled. The significant issue is not the recompilation cost, but that the design has prematurely frozen an important parameter.

The Ada 9X solution allows the introduction of arbitrary alarm levels by using a protected type Alert, and an **abstract** handling procedure Handle as follows:

```
package Base_Alert_System is

   type Alert is tagged null record;

   procedure Handle(A: in out Alert) is <>;

end Base_Alert_System;
```

Our base system requires no data with the Alert and hence the record is empty (explicit in Ada, as usual). The notation <> indicates that the subprogram is abstract and that a body will be provided with instances thus:

```
with Calendar;
with Base_Alert_System;
package Normal_Alert_System is

   type Low_Alert is new Base_Alert_System.Alert with
      record
         Time_Of_Arrival: Calendar.Time;
         Message: Text;
      end record;

   procedure Handle(LA: in Low_Alert);

end Normal_Alert_System;
```

The Normal_Alert_System refines the base system by adding the required data to the tagged type and provides appropriate instances for the Handle abstract operation. The advantage of this approach is the ease with which a new alarm level can be implemented without touching the existing code:

```
with Calendar;
with Normal_Alert_System;
package High_Alert_System is

   type High_Alert is new Normal_Alert_System.Low_Alert with
      record
         Ring_Alarm_At: Calendar;
      end record;

   procedure Handle(HA: in High_Alert);

end High_Alert_System;
```

The bodies of the various Handle procedures are not given here, nor the additional data types that such bodies might require. The important point is that adding data to tagged types can provide the same simplicity of program extension afforded by adding procedures.

There is one aspect of this approach which needs to be considered carefully in the context of critical systems. The choice of the actual body of the procedure Handle will depend, in general, upon the tag associated with the data. This will sometimes imply dynamic **dispatching** of operations which will make static analysis of programs more complex.

3.1.3 Program Libraries

The Ada library mechanism is the key to the construction of large systems since it provides security of the integration process. In Ada 83, there are problems with very large systems due to the flat, single level name space for library units. In practice, vendors who provide high-level support of large system construction have added features to overcome this problem.

In Ada 9X, library units may have **child library units**, so that an entire hierarchy can be built. Adding a child unit is established without textual change so that compilation costs are reduced compared with Ada 83.

Also, a program can be divided into partitions, in which the partitions can run in separate address spaces, again reducing the risk of monolithic development.

3.1.4 Tasking

Ada 9X introduces protected types. These types are composite and provide synchronized access to their inner components via a number of protected operations. Objects of protected types are passive and do not have a distinct thread of control; the mutual exclusion is provided automatically. The construction of systems using the Mascot principles will be very easy with protected types.

The protected types provide another communication paradigm which is less abstract and lower level than that of the rendezvous. Combined with other low-level features in 9X, one can more easily interface with operating systems and other software that works on different principles than the Ada 83 rendezvous.

One common requirement in the tasking and also the security context is to have very precise control on data. 9X introduces **controlled types** in which initialization and finalization of objects of the type is provided (constructors and destructors in C++). Types such as a Token can therefore be easily implemented as a controlled type.

3.2 Safety issues

It was recognised that the specific concerns of safety and security can run counter to others, such as those requiring the fastest possible execution. Hence it was decided to include an optional Annex within the standard to meet the requirements of safety and security. Users with these concerns can therefore decide to use an implementation that supports the Annex, rather than specify a special set of facilities (the current practice which is very expensive since a custom-built compiler is required).

The Ada 9X Annex does not introduce lots of new language features, since the key issue is assurance rather than functionality. The major parts of the Annex are as follows:

Pragma control. The user can control the entire program (actually a partition) by means of a pragma. For instance, the user can state that tasking is not used, thus allowing the run-time system to be very much smaller and simpler. Similarly, the user can ensure that other language features are avoided. By intelligent use of this facility, the language used on a particular system can be reduced to a small subset of Ada 9X. This can facilitate the analysis of the source code, since 'dangerous' language features can not only be avoided, but one can be assured that these restrictions are enforced by the compiler. (Note that IDS 00-55 [5] requires that subset restrictions are enforced by tools.) A compiler must still be able to support the entire language if the pragmas enforcing restrictions are not used.

One can argue that one would be better off with a smaller language than Ada 9X for safety-critical systems, rather than providing a mechanism for subsetting 9X. However, agreement on a suitable subset is difficult since specific applications do require every feature in 9X for convenient and maintainable coding.

Predictable execution. By invoking the pragmas which state that the Annex is to be enforced, a user is implicitly requesting predictable execution. This in turn implies documentation from the vendor of the effect of situations which would otherwise be unspecified. To reduce the impact of an unspecified part of the language, a user can request that scalar values are initialised with an out-of-range value. In essence, one would like the vendor to document the semantics of all programs that compile, not just those which adhere to the requirements of the standard (ie, no grey area in Figure 2). In practice, this is only possible if some restrictions are imposed upon the language facilities used. As an example of the problems, an Ada task can be aborted asynchronously. If this is done, and the aborted task was executing an assignment statement of an object of a complex type, then the store may be in an inconsistent state, resulting in unpredictable execution. In consequence, in order to adhere to the requirement of predictable execution, the user must not use certain features (such as the abort statement), and the vendor may enforce these restrictions.

Reviewable object code. This is an implementation requirement to provide information about the object code to permit independent analysis of the program for assurance purposes. In practice, this allows the validator to relate the object code to the source code. Since compilers cannot be guaranteed to be correct, the most critical applications do require that checks are made on the object code. Again, the vendor may impose restrictions in order to provide simpler documentation and perhaps to remove some forms of code optimization.

Inspection points. Points in the program text can be marked as inspection points. At the corresponding points in the object code, the vendor is required to provide means of determining the values of nominated objects. This implies that the object code can be analysed with special tools so that properties of the code and object values can be verified, independently of the source code. In theory, full mathematical verification could be undertaken, although this implies that the specification of the application is available in a suitable form.

Validation of values. This is a simple language feature which allows the user to enquire if the bit-pattern for an object is a legal value. For instance, if a record is read from a file written by a COBOL program, it is important that the

Ada program can check that the value is meaningful rather than execute code based upon the premise that the value is legal. In the context of safety-critical applications, such alien data is likely to be provided by some external device.

These facilities facilitate the production of safety systems of varying degrees of criticality. The user can get compiler support to place restrictions on the Ada constructs used. It will then be easier for other tools to analyse the code for critical properties.

4   Conclusions

Ada 83 is already one of the best languages available for writing critical systems as can be seen from its use on military systems like the European Fighter Aircraft, and civil ones such as the signalling in the Channel Tunnel. The enhancements made as part of the Ada 9X project should increase the ability to validate the program, and improve the visibility of the object code (the truly critical output from the software process).

5   Acknowledgement

Some of the examples have been taken from John Barnes' introduction to Ada 9X, with permission of Intermetrics, the company contracted by the US Department of Defense to revise the Ada language. Offer Pazy provided valuable comments on an early draft of this paper.

References

[1] J G P Barnes, Highlights of Ada 9X. Ada Yearbook 1993, edited by Chris Loftus. IOS Press, 1993, pp217-249. (Contains other useful information about Ada.)

[2] B A Carré and T J Jennings. SPARK — The SPADE Ada Kernel. University of Southampton. March 1988.

[3] W J Cullyer, S J Goodenough and B A Wichmann, "The Choice of Computer Languages in Safety-Critical Systems", Software Engineering Journal. Vol 6, No 2, pp51-58. March 1991.

[4] Guidelines on Risk Issues. The Engineering Council. February 1993. ISBN 0-9516611-7-5. (Free to Chartered Engineers.)

[5] Interim Defence Standard 00-55, "The Procurement of Safety Critical Software in Defence Equipment", Ministry of Defence, (Part1: Requirements; Part2: Guidance). April 1991. (Reproduced in 'Software in Safety-Related Systems', Edited by B A Wichmann, and published by Wiley and the BCS.)

[6] Ada 9X Project Report — Ada 9X Requirements. Department of Defense. December 1990.

[7] IEC/SC65A/(Secretariat 122) "Software for computers in the application of industrial safety-related systems". Draft for comment, December 1991.

[8] Issued in the USA by the Requirements and Technical Concepts for Aviation (document RTCA SC167/D0-178B) and in Europe by the European Organization for Civil Aviation Electronics (EUROCAE document ED-12B).

[9] B A Wichmann. Insecurities in the Ada programming language. NPL Report DITC 137/89, January 1989.

USING FORMAL TRANSFORMATIONS FOR THE REVERSE ENGINEERING OF REAL-TIME SAFETY CRITICAL SYSTEMS

K H BENNETT AND M P WARD
COMPUTER SCIENCE DIVISION
SCHOOL OF ENGINEERING & COMPUTER SCIENCE
UNIVERSITY OF DURHAM
SOUTH ROAD
DURHAM
DH1 3LE

Abstract

A practical tool is described which enables the user to extract high level specifications from existing source codes, using semantic preserving formal transformations. A brief overview of the theoretical foundation is given. Extensions are then described to support the acquisition of explicit timing information from real-time source codes.

Indications are then given of how the approach may be used to study safety critical properties of existing real time systems used in the process control industry.

1. Introduction

A number of companies are faced with the problem of justifying that existing code conforms to safety critical standards. A typical example is the real-time software system controlling a manufacturing process.

One alternative would be to discard the existing code completely, and start again with the additional requirement of meeting statutory or regulatory safety critical requirements. Unfortunately, this is a very expensive option. Additionally, the existing code, however imperfect, represents years of accumulated experience and refinement which (although difficult to calculate) represents a valuable asset.

There is then a motivation to examine the existing code, to explore its safety critical properties. The project described in this paper addresses one aspect of this problem, the extraction of high level specifications from real-time code, in order to:

a.   represent the code in a form more appropriate to human comprehension.

b.   determine real-time properties explicitly, based on implicit information in the source codes.

It is of course recognised that the presentation of safety critical arguments involves wider issues than those above.

The problem we address is thus strongly related to the reverse engineering problem in software maintenance. For the past nine years, we have been undertaking research in formal transformation systems; a transformation is an operation on a program which preserves or refines its semantics but converts it to another form. We have used this approach both to develop efficient executable code from high level requirements specifications, and also to achieve the acquisition of such specifications from "legacy software" given only the source codes.

This paper describes our results in this area. We discuss the reverse engineering problem in software maintenance, and show how we have addressed this within the 'Maintainer's Assistant' project. This project addressed only sequential code, but in a more recent research project, we have developed novel techniques for analysing real time systems which are based on a shared memory model of concurrent computation. This has been applied both to conventional real-time programs and to Programmable Logic Controllers (PLCs). Finally, we indicate the directions we are taking in using this information in safety critical systems.

2.   Transformation Systems for Maintenance

Many methods have been suggested which address the problem of producing software in such a way that it performs as expected, is not delivered late and is not over budget. Wirth [1] proposed one solution in which a program is developed incrementally by *stepwise refinement*. However, the problem still remains that each step is done intuitively and must be validated to ensure that it preserves the correctness of the program with respect to some specification. An improvement on stepwise refinement is to allow only provably semantic-preserving changes to the program. Such changes are called **transformations**. Program development by transformation offers many advantages over *ad hoc* program development methods and these advantages can be summarised as follows [2] [3]:

- The final program is correct (according to the initial specification) *by construction*.

- Transformations can be described by *semantic rules* and can thus (as can programming languages) be used for a whole class of problems and situations and are not restricted to a particular type of program.

- Because of the formal, clerical, nature of program transformation, the whole process of program development can be supported by the computer. Performing the many small changes required by transformations manually would almost inevitably introduce clerical errors and the situation would be no better than the original *ad hoc* methods. However, such clerical work is ideally suited to automation, and this allows the computer to carry out the monotonous part of the work while the programmer concentrates on the actual design decisions.

- The approach is quite flexible in that the overall program structure is no longer fixed throughout the development. Major changes can be made with the knowledge that the program's functionality will remain unchanged.

Much research has been carried out on transformation for program development and summaries can be found in papers by Partsch and Steinbrugen [4], Feather [5] and Yang [6]. However, transformation for software maintenance has not been so comprehensively addressed. One method of using transformations to perform maintenance, suggested by Balzer [11], is to modify the original specification and then reimplement it. This has the advantages that, in the specification, information is localized and loosely coupled, whereas, after implementation, the information is reflected throughout the program and many of the connections are merely implicit.

Many existing programs only exist in the form of source code which has often been heavily modified and extended over years of software maintenance. If documentation or high level design information does exist, it is typically inconsistent with the source code and cannot be relied upon as an accurate representation. When the code is the only dependable resource, we have nothing we can edit and re-transform to produce a new version of the program. The requirement of being able to maintain code based solely on the source code was the motivation behind the production of the **Maintainer's Assistant** by the University of Durham, IBM UK Ltd. and Durham Software Engineering Ltd. in the **ReForm** project.

The Maintainer's Assistant, which has been described in detail in previous papers [7] [8] [3] [10], is a code analysis tool aimed at helping the maintenance programmer to understand and modify a given program. Program transformation techniques are employed by the Maintainer's Assistant both to derive a specification from a section of code and to transform a section of code into a semantically equivalent form. The Maintainer's Assistant has at its core a Wide Spectrum Language (WSL). All transformations are expressed in terms of WSL constructs, and all proofs of transformation correctness are based on WSL semantics (note however that the user of the tool is not concerned with undertaking such proofs).

3.   The Process of Program Transformation

The ultimate objective of the Maintainer's Assistant is to facilitate the transformation of existing, large-scale source programs to high level requirement specifications.   It is envisaged that such specifications will, in general, be represented in non-executable form, using a language such as Z or VDM. The major attractions of this approach are that the specification will be semantically equivalent to the original code or the latter will be a refinement of the specification. Thus the user can be confident that the specification can provide a representation which can be maintained in place of the original source code. Maintaining a high level, more abstract representation has a number of important advantages:

- it is more compact than the source code;

- it is expressed in a more problem-oriented notation;

- executable code can potentially be generated from it automatically or semi-automatically.

It is recognised that the acquisition of a specification at a high level of abstraction cannot be an automatic task; this problem is undecidable.

The process adopted in the Maintainer's Assistant is as follows:

1.   The source code is restructured to provide an easily understood representation by improving control flow, removing dead code and, importantly, introducing a good procedural structure. Transformations are used to accomplish this restructuring.

2.   The restructured source code is then progressively transformed, by the use of procedural and data abstraction, to a high level requirement specification.

3.   Appropriate abstractions are identified and introduced.   For this the maintainer uses expertise of the application domain and software engineering to provide a strategy.

At low levels of abstraction, much of the transformational activity is clerical and can be substantially automated. At higher levels, the expertise of the maintainer is required. Hence, the Maintainer's Assistant is fundamentally an interactive tool.

In this paper, the results of addressing the first step in the process are presented for real programs of medium size (up to 20,000 lines of source code, comprising heavily modified, geriatric Assembler).

Although this first stage may appear to be little more than conventional automatic control flow restructuring, the traditional automatic tools do not provide a good basis for subsequent abstraction transformations. The shortcoming of such tools have been pointed out by [9]:

1.   they may replace complex control flow with complex data flow, by introducing flag variables;

2.   they may result in programs that are considerably larger;

3.   they do not help human understanding of the system.

In the Maintainer's Assistant, interaction is used to restructure the program, thus avoiding the above problems and providing an appropriate starting point for subsequent abstraction transformations. For example, the Maintainer's Assistant supports the identification of suitable code sections to fold into procedures; typically, a large monolithic unstructured program can readily be transformed into a short main block from which calls to a set of sub-procedures are made. We are not aware of automatic restructurers which can achieve this but it *is* an important intermediate step in identifying abstract data types.

4.   Major features of the Maintainer's Assistant

The Maintainer's Assistant (Figure 1) includes the following major features:

• It acts, initially on existing program code, as a tool to aid comprehension by producing well-structured equivalents which reflect the specification of the system.

• Only the program code is required.

• The system can, potentially, work with any imperative language by first translating it - with a stand-alone translator - into the system's internal language. Currently these front-end translators only exist for IBM 370 Assembler and sub-sets of Basic and Pascal.

• The internal language, WSL, is a wide spectrum language. This language can be used equally well to express non-executable specification as to express programs. The language is mathematically based so that proving the correctness of a transformation amounts to proving the equivalence of two formulae. This contrasts with C, for example, in which transformations are difficult to discover and prove. WSL is described in detail in the papers by Ward, [12] [13] [14] [15] [16].

- Transformations are themselves coded in an extension of WSL called **Meta-WSL**, making maintenance of the system much simpler.

- The system incorporates a large (but easily accessed) catalogue of transformations, some of which will be described later.

- The applicability of each transformation is tested before applying it, unlike in some systems which rely on the user to do this.

- A history/future structure is built in to permit back-tracking and forward-tracking allowing the programmer to change his or her mind.

- The system, as was stated earlier, needs to be interactive and thus incorporates a front end and pretty-printer called the **Browser** [10]. This is built on X-Windows and uses the Motif window manager.

- The system includes a database structure to store information about the program being transformed. Typical of this information would be the variables that are used in a particular section of code [10].

- The system incorporates a symbolic simplifier for mathematical and logical expressions.

- The system also includes a method for calculating metrics for the code being transformed.

- The transformation engine which forms the core of the tool is built on the principle of having a hierarchy of *abstract machines*.

5. The Transformations Needed for Real Code

Transformation systems rely on a predefined collection of rules which define how to change the program *while still preserving its semantics* and the Maintainer's Assistant is no exception. Each rule can be categorised as one of the following types:

- Rules which can be expressed by means of pattern-matching and replacement. An example is the reordering of a conditional statement by negating its test condition.

- Procedural, algorithmic transformations such as removing a dummy loop.

- Hybrid rules which are a combination of the above. Burstall and Darlington's fold/unfold rule is of this type [17].

Despite the fact that all approaches will include these categories of rules, there are, in general, two (not necessarily contradictory) ways of constructing the collection of rules.

The first method is the **catalogue approach**. In this approach there is a large set of transformations covering all aspects of program development. For example, there could be rules relating to programming knowledge, features of the language, the programming domain, efficiency of implementation and choice of data structure.

This method, although powerful, has two drawbacks. Firstly, the rules are fixed so that if the system is used outside its perceived domain, it becomes less suitable and secondly, with such a large catalogue, finding the "best" transformation to apply at a particular point is relatively time consuming, especially if the programmer is unfamiliar with all the options at his or her disposal.

The other method is the **generative set approach**. In this approach there is a small set of powerful, language-independent, elementary transformations from which others can be produced by combination.

Compared with the catalogue approach, this method is much more flexible since transformations appropriate to the situation can be constructed by the programmer from sequences of the elementary transformations. This advantage is also the drawback, however, since the programmer's effort is shifted to trying to work out what sequence of very minor changes he or she needs to do in order to produce some desired large scale effect.

The Maintainer's Assistant would appear to fall into the former category, since it has over 600 transformations in its catalogue. Although this might seem to be far too many for a user to become familiar with, different programs do require different types of transformation and, more importantly, having that many allows individual users to work in their own preferred way with their own preferred set of transformations, as described in the next section.

5.1   Elementary Transformations

The Maintainer's Assistant includes all the elementary transformations (such as inserting assertions) that have been proved by Ward [18]. These correspond to the transformations of the generative set approach. The maintainer may choose to use sequences of these simple transformations to accomplish some more complex effect, such as changing the representation of a data structure. If this is how he or she prefers to use the system then there are only a relatively small number of transformations with which to become familiar.

210

5.2 Compound Transformations

In using the Maintainer's Assistant, a number of common sequences and combinations of transformations have been identified by experience and case studies. Rather than expecting the maintainer to remember such sequences, they have been built in as transformations that can be selected in the same way as the elementary transformations. Although there are a large number of these compound transformations, the efficiency gained by learning to use even just a few of them seems to outweigh the initial learning time.

5.3 Generic Transformations

A great many of the elementary (and some of the compound) transformations are variations on themes. For example, there are separate elementary transformations for taking a statement out of a local variable structure and for taking a statement out of a loop. In addition to the many distinct transformations, these themes have been combined into single generic transformations so that the user does not need to know which specific transformation has to be selected. Thus, a maintainer who adopts this method of working can accomplish much of his or her work using only a very small number of transformations.

Although many operations can be classed as one of the twenty generic transformations, some operations cannot be incorporated into the generic set since the user has to be more explicit in stating the desired form of the transformed program. For example the user has to be able to choose between changing an unbounded loop into a "For" loop or into a "While" loop.

5.4 Selecting Transformations from Menus

The maintainer uses the system by first pointing to a piece of code and then selecting a particular category of transformation. These categories are "move", "join", "insert", "delete", "reorder", "rewrite", "use/apply", "simplify/delete", "multiple" and "complex". The system then tests the applicability of all the relevant transformations of this type and presents the user with a menu of all the valid ones.

The transformations have been subdivided further into the "elementary", "compound" and "generic" groupings defined above. The menus only display transformations belonging to the groupings that the user has selected and are, therefore, not unduly cluttered.

6.   The Design of a Practical Tool

The introduction stated that the main contribution of this paper is to identify those features of the Maintainer's Assistant that, from practical experience, are essential to the design of a usable tool for transformation-based reverse engineering. The experience has been gained by applying the tool to a series of real Assembler programs (not "toy" examples). These Assembler programs are typical of heavily-maintained, unstructured, medium-scale programs (of up to 20,000 lines).

It has already been stressed that the Maintainer's Assistant is fundamentally interactive in operation, even though at a lower level of abstraction many clerical or common tasks can be automated using complex transformations. It has also been explained that a combination of the generative and the catalogue approaches addresses the needs of a wide range of program types and, in addition, helps the new user who, when learning about the system, may wish to concentrate on a few key transformations.

6.1   Design of the Tool

A practical system for reverse engineering has to do with real programs, not laboratory or toy examples. More specifically, the following requirements are identified for the tool at the outset of the project:

*   the tool must cope with all the usual programming constructs and their uses and abuses, including go tos, global variables, aliasing, recursion, pointers, side effects etc.

*   it is not acceptable to assume that the code has been developed or maintained using structured methods. Real code must be acceptable and major restructuring may be required before proper reverse engineering can start. Such restructuring should be carried out as far as possible automatically by the system.

*   the transformations in the library must be proven correct, so that the user can employ them with confidence, but also so that the user does not have to undertake such proofs. The transformation must meet applicability conditions, so that these can be mechanically checked by the tool.

*   the correctness of the implementation must be established.

*   it must be possible to select a sub component of a large existing system and to guarantee to preserve the interactions of the sub component with the rest of the system. This permits attention to maintenance hot spots.

6.2 Tool Architecture

The main components of the tool are shown in Figure 1. The core of the tool is a library of proven transformations, together with the transformation engine. The transformations in the library are proven before the tool is built. They allow construct in WSL to be recast into another WSL construct while ensuring that semantics are preserved. Thus, the user has only to select a transformation and apply it. He or she does not have to do the proof, and the systems transformation engine checks that the transformation is applicable.

As noted previously, the first stage is to load the source code into the tool, and this is achieved by a front end translator. The equivalent WSL is stored internally as an abstract syntax tree, together with ancillary information to aid applicability checking. The user views the WSL through a browser interface and X Windows front end. A typical display is shown in Figure 2.

The protocol between the browser and the front end is very simple, and is represented in ASCII. This means that the front end can easily be installed on a different machine from that running the core tool. We have demonstrated this in practice by running the front end on a PC, a PS/2, a Macintosh, and a UNIX workstation, while the main tool has run on an RS 6000 workstation and a SUN workstation.

The WSL program is presented in a main window in a pretty printed form convenient for the user. The WSL in Figure 2 was produced from an IBM System 370 Assembler program.

The system is inherently interactive, and the two main ways of interacting with it are:

(a)   Selecting a WSL construct to transform.

(b)   Selecting a transformation from the library to apply next.

Buttons are used to control both interactions. The selection of a WSL construct is syntax directed so only valid syntactic units can be chosen. The construct selected is displayed in reverse video. As shown in Figure 2, the transformations are grouped into 10 categories, controlled by buttons, in order to simplify selection from the library of over 600. The system automatically checks the applicability conditions of each transformation against the selected WSL construct. Thus when a user selects a category of transformations by clicking on a button, the resultant pop up menu only contains those transformations if any that are valid. We have found through experience that a user often employs a regular pattern of

213

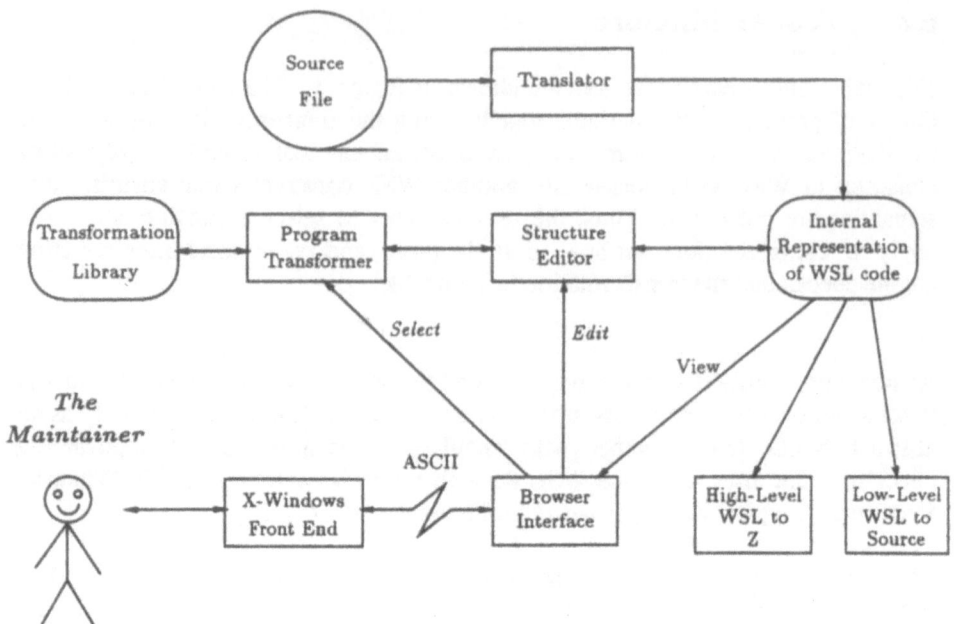

Figure 1: A Block Diagram of the Prototype Tool

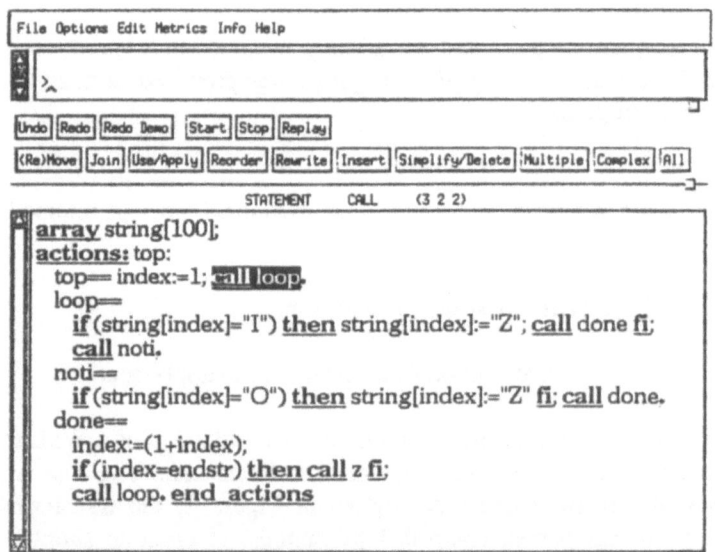

Figure 2: A screen dump from the Prototype Tool

transformations and it is easy within the tool to group such transformations at a more powerful composite single transformations. Currently such transformations are handled by the tool builder, but as transformations are represented internally in WSL, it would not be difficult to allow the user to do this.

The system is constructed as a hierarchy of abstract machines, each of which is formally specified. The architecture is shown in Figure 3.

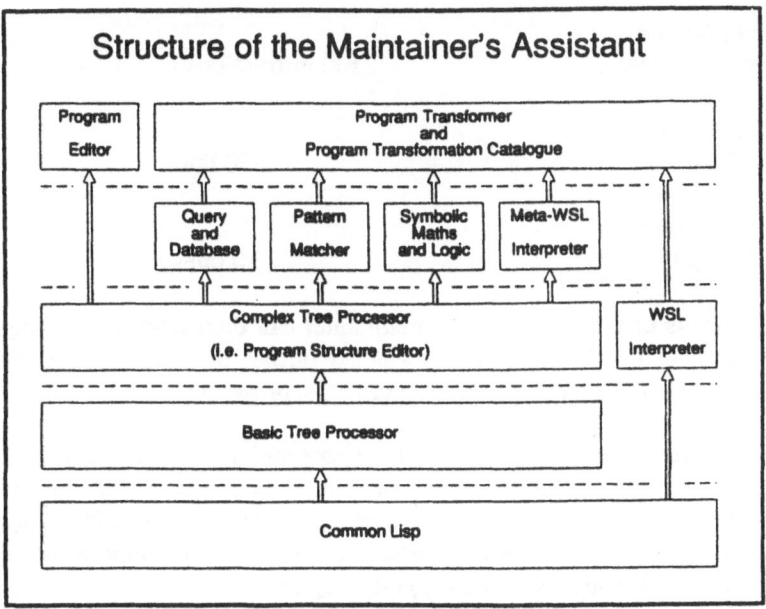

Figure 3 Structure of the Maintainer's Assistant

7.   Transforming Real Code

The Maintainer's Assistant has been produced as part of a project which has been partially funded by IBM UK Laboratories and by the UK Department of Trade and Industry and SERC. Part of the project has involved taking source code written in IBM 370 Assembler and using the system on it.

In order to do this, it is necessary first to translate the Assembler in WSL and this is done by means of an automatic translator. The translation is very simple-minded. Each instruction of assembly language maps into several lines of WSL because the WSL has to reflect every aspect of the semantics of the instruction, such as the setting of flags, even if these implicit aspects are not always needed. In this way, it is possible to simplify the verification of the implementation of the Assembler-to-WSL translator.

215

Once the code has been translated into WSL, it is then possible to use the Maintainer's Assistant to remove all the extra, unnecessary code that was introduced by the translation process and to do some simple tidying of the code. A powerful compound transformation has been written to perform this initial simplification. When performed on a WSL program that has been translated from Assembler, it reduces its size considerably. The figures below give an indication of how the program changes during these processes.

Assembler program
before the translation   100 instructions

WSL program after
the translation   250 lines

WSL program after
the initial simplification   80 lines

Once the code is in a tidier form, the maintainer can then work on it, simplifying and restructuring it. Doing this has yielded further improvements to the code through the identification and production of loops, procedures and data structures.

The tool has been used on a selection of systems and applications written in IBM 370 Assembler. In all instances, a maintainer who was unfamiliar with the particular modules was responsible for working on the code and in all instances very significant improvements were made, not only in the structure of the code but also in its comprehensibility. The longest program module on which the system has currently been used is about 20,000 lines of WSL.

As a result of these experiments, the following have been identified as important features and criteria for a successful tool based on formal program transformations:

- It is essential for the tool to be fundamentally interactive, to allow human expertise of both software engineering and maintenance and of the application domain, to influence the direction of the transformation process.

- The tool needs to work in a windows environment together with some important supporting features: a clear definition of transformation types (there are ten in the Maintainer's Assistant); a forward/backward history mechanism; automatic testing of transformation applicability; structural, syntax-based editing for when errors are found in source programs; facilities to "fold" and "unfold" program sections so that large programs can be shown on a small screen.

- The use of both generative set and catalogue approaches and the inclusion of very complex, powerful transformations provides the ability to address all situations and the means to harness human expertise.

- It is essential for the tool to be used with confidence by the maintainer. In the ReForm project the transformations have already been proved to be correct [18], so all that remains is to demonstrate the correctness of the implementation; something which could be done by program transformation.

- The provision of metrics give the user guidance whilst providing management with a measure of benefit.

- Automation of low level detail saves the programmer from the formidable task of program understanding until a higher level of abstraction is achieved.

- A method for using the tool is important.

- A user guide and online help should be provided.

8.   Modelling Interrupt-Driven Programs in WSL

WSL has no notations for parallel execution or interrupts. We chose not to add such notations to the language, since this would complicate the semantics enormously and render virtually all our existing transformations invalid. Consider, for example, the simple transformation:

$x := 1;$ **if** $x = 1$ **then** $y := 0$ **fi** $\approx x := 1; y := 0$

which is trivial to prove correct in WSL (the notation $\approx$ means 'equivalent to'). However, this transformation is not universally valid if interrupts or parallel execution are possible, since an interrupting program could change the value of x between the assignment and the test.   Instead, our approach is to *model* the interrupts in WSL by inserting a procedure call at all the points where the program could be interrupted. This procedure tests if an interrupt did actually occur, and if so it executes the interrupt routine, otherwise it does nothing. Although this increases the program size somewhat, the resulting program is written in pure WSL and all our transformations can be applied to it.

One of our aims in transforming the resulting WSL program is to move the interrupt calls through the body of the program, and collect them together and merge them as far as possible using transformations to do this. The body of the main program would then be essentially sequences of statements from the original program, separated by the processing of any interrupts which occurred during their execution.

In order to model time within WSL we add a variable *time* to the program which is incremented appropriately whenever an operation is carried out which takes some time. We can then reason about the response times of the program by observing the initial and final values of this variable. We can also model the times when interrupts occur by providing an input sequence consisting of *pairs* of values <t, c> where *t* the time at which the interrupt occurs and *c* is the character. Naturally we should insist that the sequence of *t* values be monotonically increasing. Such a sequence can model input from an external device, a concurrent process, or even a hardware register. We make this explicit in our model of the program: the array (or equivalently, sequence) *input* consists of pairs of times and characters to represent the inputs, and is sorted by times. The interrupt routine tests the *time* variable against the time value associated with the first element of the input sequence to see if that interrupt is now "due". If so, then it removes a pair from the head of the sequence and processes the result. The program is modelled as follows:

S_1;   → S_1;
S_2 *interrupt(time); time := time + 1;*
etc ... S_2;
                                    interrupt(time); time := time + 1;
                                    etc ...

If we assume a discrete model of time, i.e. that the value of *time* is an integer, and we assume that time is incremented by one between each potential interrupt, then the test for validity of a call to the interrupt routine is simply:

interrupt(time$_0$)   →   **if**$(time_0 = input[1][1])$
                                              then *process_interrupt* **fi**

where *process_interrupt* corresponds to the original interrupt service routine. Note that input[1] is the first element of the input sequence: this element is a pair of values(a time and a character); so *input*[1][1] is the first element of the pair; i.e. the time of the first interrupt.

However, a better model, which does not require a discrete model of time, and which allows different "atomic" (i.e. non-interruptable) operations to take different amounts of time, is the following:

interrupt(time$_0$)   →   **while**$(time_0 >= input[1][1])$ **do**
                                          process_interrupt **od**

This revised model of the interrupt routine allows more than one interrupt to occur between atomic operations, and has the advantage that a call to *interrupt* can be merged with a second call which immediately follows it:

interrupt(t_1); *interrupt*(t_2) $\approx$ *interrupt*(t_2)

provided $t_2 >= t_1$. This follows from the transformation:

while B_1 **do** S **od**; $\approx$ **while** B_2 **do** S **od**
while B_2 **do** S **od**

provided $B1 \Rightarrow B2$. This transformation is proved in [9]. If $t_2 >= t_1$ then we have $(t_1 >= input[1][1]) \Rightarrow (t_2 >= input[1][1])$, and we can merge the two **while** loops and hence the two procedures.

Thus, once we have moved a set of interrupt procedure calls to the same place, we can merge them into one statement equivalent to "process all outstanding interrupts", which is much closer to a specification level statement than is a series of calls to the same procedure.

Note that the addition of interrupt calls is defining the "interruptable points" in the program, or equivalently, the "atomic operations". The increments to *time* define the processing time for each atomic operation. For real programming languages, e.g. Coral, the atomic operations may well be machine code instructions, rather than high-level language statements, and it is at the machine code level that the model needs to be constructed, for it accurately to reflect the real program. This will inevitably lead to a large and complex WSL program; however automatic restructuring and simplifying transformations can eliminate much of the complexity before the maintainer even has to look at the program.

9. Concurrency

Interrupts may be regarded as a special type of concurrent processing on a single processor. When an interrupt occurs, the "main" program is suspended and the interrupt routine is executed in its entirety, possible changing the state of the main program in the process. Execution of the main program then resumes. It is the fact that the interrupt routine is executed in its entirity that makes interrupts a special case from the point of view of modelling in WSL; we are able to insert a copy of the interrupt routine wherever an interrupt occurs, and hence the effect of the interrupt is deterministic.

The analogy with a single-processor multitasking system is obvious: here the running program executes until it is suspended by the operating system. Other tasks are then (partially) executed, and may change the state of the original program; eventually execution of this program is resumed. From the perspective of the original task, this looks like a call to a procedure which executes sequences of instructions from the operating system and the other tasks in the system, and subsequently returns. The analogy also applies to more general forms of concurrency, including fully-parallel multiprocessor systems. In principle, the state of any program or task in such a system may be changed, between one atomic operation and the next, by other concurrently executing tasks. Again this could be modelled by procedure calls between each pair of atomic operations, which perform the appropriate processing and change the state accordingly.

In our experiments, we are therefore attempting to model a concurrent system in WSL by modelling the potential interaction between tasks. For each task, we derive a specification for the conditions on its state which are guaranteed to be preserved by the other tasks in the system. This is done by looking at each of the remaining tasks in isolation, and deriving for each of these a specification on the conditions of the state of the first task which are guaranteed to be preserved by the execution of any instruction in the second task. We thus have derived the condition on the state of the first task for non interference by the second task. We can now repeat this process for all the remaining tasks, and then combine, using non deterministic choice, to give us an overall specification of all the conditions on the state of the first task guaranteed to be preserved by the execution of any instruction from any other task in the system. The full specification for the effect of the remaining tasks on the first is then given by any number of concatenations of this specification. We now proceed in an analogous way to that used for interrupts. We substitute this specification between each atomic action in the task. This gives an abstraction of the task itself including the effect of interference by the rest of the system. We are now in a position to reverse engineer this model of the task, using our existing transformations, in order to find a specification for the task incorporating the effect of interference from other tasks.

We can repeat this process for each of the tasks in the systems, and having derived a specification for each them, we can obtain the specification for the complete system by combining the individual specifications using the WSL join operation.

Our modelling approach for both interrupt and concurrent systems has been used in a series of simple examples: some of these have been taken from real applications.

This study shows that by using an appropriate models we can represent interrupt-driven and concurrent programs within the (purely sequential) WSL language. With such models, reverse engineering techniques of [16] can be applied to extract the specification of the original program. Program transformations are sufficiently powerful to cope with these, often complex, models. Although a fairly large number of transformations are required to deal with these models, results from case studies indicate that these are used in a systematic way: this suggests that much of the work can be automated by a tool such as the Maintainer's Assistant and this is currently being further investigated.

10. Safety Critical Systems

Over the next three years we plan to apply and extend these techniques as an aid to the comprehension of existing safety critical programs. After the event evaluation of geriatric systems does of course involve much wider issues than extracting specifications. However, it is hoped that the techniques for dealing with concurrent and shared memory parallelism of real time programs described in this paper can be used to extend an existing tool, the Maintainer's Assistant, to help the process.

11. Results

The Maintainer's Assistant project has demonstrated that a tool to assist reverse engineering of sequential systems, based on formal transformation systems, is feasible for non trivial industrial scale problems. The tool has been used with IBM 370 Assembler to date and while this is of minority interest, it is also an extremely challenging example. We have demonstrated that a more comprehensible form of an existing heavily maintained program can be obtained relatively easily using the Maintainer's Assistant.

More recent theoretical work has suggested that the feasibility of modelling concurrent programs in such a way that existing transformations can be applied to them to obtain high level specifications. It is planned to study this further, and to extend the Maintainer's Assistant tool, over the next few years.

12. Conclusions

The Maintainer's Assistant is based on a strong theoretical line of research which predated the practical tool. The development of the Wide Spectrum Language WSL is seen as the crucial foundation for the work. It provides a framework for proving a wide range of powerful transformations. The underpinning of WSL by infinitary logic has allowed us to prove powerful transformations between iterative and recursive constructs.

Our approach in addressing concurrency has been to add to this theory without invalidating results to date. Indications so far are that we can model concurrent programs by explicit insertion of 'worst case' interaction. followed (using existing transformations) by a process of restructuring to simplify the program.

For safety critical systems, our aim is to develop this approach to help the user comprehend the program, even if it has been in service and heavily maintained over a period of years.

13. Acknowledgements

The research described in this paper has been funded partly by a Department of Trade and Industry SMART award to the Centre for Software Maintenance Ltd., and partly by SERC (The Science and Engineering Research Council) project "A Proof Theory for Program Refinement and Equivalence: Extensions". We particularly acknowledge the contribution of Edward Younger in the research on concurrency. Funding by the UK DTI and SERC for the Maintainer's Assistant is acknowledged. We also wish to thank IBM UK Ltd for their funding, and for their technical contribution to and support of the project.

References

1.   Wirth N., "Program Development by Stepwise Refinement", Comm ACM, vol 14, 1971, pp 221-227.

2.   Bauer F.L., Moller B., Partsch H. and Pepper P., "Formal Construction by Transformation - Computer Aided Intuition Guided Programming", IEEE Trans Software Eng, 15, (Feb 1989).

3.   Bull T., "An Introduction to the WSL Program Transformer", Conference on Software Maintenance, 26-29 November 1990, San Diego, (Nov 1990).

4.   Partsch H., "The CIP Transformation System", in Program Transformation and Programming Environments, Report on a Workshop directed by Bauer F.L. and Remus H., Pepper P., ed, Springer-Verlag, New York-Heidelberg-Berlin, 1984, 305-323.

5.   Feather M.S., "Specification Evolution and Program Re(transformation)", Proc 5th RADC Knowledge-based Software Assistant Conf, (Sept 1990).

6.   Yang H., "How does the Maintainer's Assistant Start?", Durham University, Technical Report, 1989.

7.   Calliss F.W., Khalil M., Munro M. and Ward M.P., "Knowledge-based System for Software Maintenance", IEEE Conf on Software Maintenance, Phoenix, Arizona, 1988.

8.   Ward M., Calliss F.W. and Munro M., "The Maintainer's Assistant", Conference on Software Maintenance, 16-19 October 1989, Miami Florida (Oct 1989).

9.   Calliss F.W., "Problems with Automatic Restructurers", SIGPLAN Notices, 23, March 1988, 13-21.

10.   Yang H., "The Supporting Environment for a Reverse Engineering System - The Maintainer's Assistant", Conference on Software Maintenance, Sorrento, Italy, December 1991.

11.   Balzer R. and Sartout W., "On the inevitable intertwining of Specifications and Implementations", in Software Specification Techniques, Addison Wesley, 1986.

12.   Ward M., "A Recursion Removal Theorem", Springer-Verlag, Proceedings of the 5th Refinement Workshop, London 8-11 January, New York-Heidelberg-Berlin, 1992.

13.   Ward M. and Bennett K.H., "A Practical Program Transformation System for Reverse Engineering", Working Conference on Reverse Engineering, May 21-23 1993, Baltimore MA, (May 1993).

14.   Ward M., "Derivation of a Sorting Algorithm", Durham University, Technical Report, 1990.

15.   Ward M., "A Recursion Removal Theorem", BCS Refinement Workshop, 8-11 January, Jan 1992.

16.   Ward M., "Abstracting a Specification from Code", Journal of Software Maintenance Research and Practice, 5, 1993, 101-122.

17.   Burstall R.M. and Darlington J., "A Transformation System for Developing Recursive Programs", J ACM, vol 24 no 1, Jan 1977, pp 44-67.

18.   Ward M., "Proving Program Refinements and Transformations", Oxford University, DPhil Thesis, 1989.

Use of Neural Computing in Multiversion Software Reliability

Derek Partridge
Noel Sharkey*
Department of Computer Science, University of Exeter
Exeter EX4 4PT, UK

Abstract

The main goal of this project is to evaluate the potential of neural computing as a diverse methodology that will enhance the reliability of software systems constructed as a set of independent versions. The first phase, however, is focused on an exploration and elucidation of a technology to support efficient and effective programming using neural computing techniques. This paper describes the results of our first experiments, and the implications for the development of a technology of neural computing as well as for the role of abstract specifications in software engineering.

1 Introduction

Multiversion software engineering uses a strategy that attempts to increase software reliability through the use of multiple, independently developed, implementations – a set of versions. The basic idea is that a set of alternative implementations (no one of which is expected to be totally correct) will provide greater overall reliability (through some majority-vote procedure, say) then any single implementation. However, empirical studies of the actual reliability obtained from multiversion software has shown, repeatedly, that the reliability gains fall far short of would be expected from the assumption that independently developed, alternative versions would fail independently – i.e. that the probability of **two** versions failing on a randomly selected input is (approximately) the **square** of the probability that one version fails on a randomly selected input. Empirical studies show that this is an erroneous assumption, and that the actual probability of two versions failing on a randomly selected input is orders of magnitude higher than the assumption anticipates [2] [1].

The presence of common design faults running through the set of alternative versions is thought to be a major reason for the lack of significant reliability gains. Therefore, it has been suggested that if alternative versions are developed using diverse methodologies the presence of common design faults would diminish, and hence multiversion reliability would increase. An empirical study using the languages Modula2 and Prolog (which force rather different design strategies on the process of implementation development) by Adams and Taha [1] provides support for this notion – methodological diversity does lead to a significant lessening of common errors.

---

*supported by research assistants, Oband Jackson and Simon Klyne

We have proposed that neural-net implementations are likely to be diverse in the extreme, and hence a multiversion approach that combined conventionally programmed and network programmed versions would further enhance overall system reliability. In addition, it has long been known in neural computing that very different network implementations can be obtained by varying network initialisation prior to training, by varying net architecture, by varying training regimes, etc. Thus we are also investigating the potential for diverse methodologies *within* the neural computing paradigm. But first the largely ad hoc art of neural computing must be better understood, it must be transformed into more of a proper technology before we can hope to exploit it effectively. Thus the first part of this research programmed is concerned largely with the problem of articulating a technology of neural-net implementations of well-defined functions, we aim to elucidate a technology of network programming.

2 Principles of network programming

Given a function to implement as a neural network, there are a number of (currently) ill-understood and unconventional stages in the process. Instead of designing an algorithm to meet the specification, we train a network to do so, and in both cases we test the implementation to determine whether it is acceptable. Currently, we are using Multilayer Perceptron nets trained with the backpropagation algorithm [8].

The stages in network programming can be set out as follows:

1. analyse the problem in order to determine which type of neural net is most appropriate, and what is the potential scope of the input and output

2. design a coding strategy for the input and output data

3. design a neural net architecture that is capable of learning the function with a minimum of resources

4. select a representative training set of input-output pairs

5. initialise the network

6. train the network to correctly compute the training set (an iterative algorithmic process)

7. test the trained network implementation

8. if unsatisfactory, analyse failed test items with respect to training set, design new training set and retrain

A lot of questions about every one of these stages can be asked, and very few of them can be answered with certainty. For example, what is a 'representative' training set for a given function? How do we analyse failed tests? And how do we design a new training set as a result of the analysis? The production of a technology can be viewed as the generation of accurate answers to most of these questions. We have begun to shed some light on some of them.

3 The Launch Interceptor problem

As a test application problem for this feasibility study, we have chosen the "Launch Interceptor Problem" (the LI problem) which is close to ideal for our initial studies (the five-page specification of this problem is reprinted in [2]). The problem is part of an anti-missile system. The input is a radar image together with 19 real-valued parameters and several small matrices which are used to control the interpretation of the radar images. The output is simply a decision Launch (when all 15 launch criteria are satisfied) or No-Launch (when one or more of the 15 launch criteria is not satisfied by the input data) based on values in a 15-bit vector. Much of the specification concerns what to compute for each of the launch criteria. These criteria are typically satisfied by the occurrance of some geometric relationship in a subset of the radar data points where the input parameters specify the constants to be used in computing the specified relationship. One criterion is, for example, three points within a circle of radius RADIUS1, where RADIUS1 is one of the input parameters. The LI problem has previously been used as an exemplar of a simple (but realistic) safety-critical problem, and has been subjected to extensive study in the context of conventional programming paradigms (e.g. [2]and [1]) which provides us with:

a) a benchmark for immediate assessment of the relative strengths of the network-programming paradigm as a model for the production of reliable software in its own right;

b) a framework within which to quantify the impact of network-programs as a contributing technology to software reliability through multiversion programming.

There is also a "gold" program (a 'certified' implementation of the LI problem which is taken as the measure of correctness in the empirical studies) has been used to generate the necessary input-output pairs for both training and testing purposes.

Input:

- a sequence of 2-100 xy coordinate points – 2 real values for each

- the number of xy points – an integer

- 19 real-valued parameters

- Logical Connectivity Matrix (LCM) – 15x15 elements each selected from the set {AND, OR, NOT_USED}

(Note that the LCM is symmetric about the diagonal, and so only 120 elements are significant.)

Output:

- Final Unlocking Vector (FUV) – 15 binary elements

- Launch or No-Launch decision

According to the specification, the 15 'geometric' functions are crucial. The complexity of the problem comes from the subleties of the general algorithms for implementing each of these functions as well as from the use of the LCM to combine the outputs from these functions (the bits of a Conditions Met Vector, CMV) to form another matrix, the Preliminary Unlocking Matrix (PUM), and then the Final Unlocking Vector (FUV).

In an earlier study ([7]), we trained networks to implement the latter half of the specification – from CMV, LCM, and PUM to the Launch/No Launch decision. We are currently exploring network implementations of the 15 functions – the Launch Interceptor Conditions, LICs.

3.1 The 15 conditions

Each of the 15 functions computes a Boolean value using *consecutive* data points. A given function will compute TRUE (i.e. output a 1 in the current network coding) if there exists *at least one* set of such consecutive points that satisfy the geometric relation specified. That is, (x_1,y_1), (x_2,y_2), (x_3,y_3) are x,y coordinate points, which can be labelled data points **A**, **B** and **C**. Having consecutive data points reflects the temporality of the input. Note that (x_1,y_1) and (x_3,y_3) would *not* constitute consecutive data points.

All of the functions can be broadly described as computing some geometric property of the trajectory formed by consecutive data points, and then comparing that value with the value of a fixed parameter. For example, there are functions that compute whether the distance between two consecutive data points is greater than the parameter LENGTH1 (Function # 1), whether the area created by three consecutive data points is greater than the parameter AREA1 (Function # 4), and so on.

Note: In any description of these functions, as the data points are *always* consecutive, the points form a sequence (i.e. have an ordering). The labelling convention $\exists$ (a,b,...p_n), indicates that there exists *at least one* unbroken sequence of data points **a**, **b** ... **p** (where the subscript n refers to a number, eg. the pth data point may be the sixth in that sequence).

- Function # 1.
  $\exists$ (a,b), where d[a,b] > LENGTH1
  The *distance* between data points **a** and **b** is greater than the value of the parameter LENGTH1.

- Function # 2.
  $\exists$ (a,b,c) $\in$ ◯ (with radius, RADIUS1)
  The three data points a, b and c cannot all be *contained* within a circle whose radius is the value of the parameter RADIUS1.

- Function # 3.
  $\exists$ (a,b,c) where $\angle abc < \pi - \epsilon$ or $\angle abc > \pi + \epsilon$, and π and ϵ are input parameters.

- Function # 4.
  $\exists$ (a,b,c), where $\triangle$ > AREA1
  The three data points a, b and c form the vertices of a triangle with an *area* greater than the value of the parameter AREA1.

227

- Function # 5.

  $\exists \, (a, b \ldots p_n) \in \oplus$.

  The 2 or more data points **a**, **b** ...p_n lie in parameter QUADS quadrants of the xy plane.

- Function # 6.

  $\exists \, ((x_i, y_i) = \mathbf{a}, (x_j, y_j) = \mathbf{b})$ where $(x_j - x_i) < 0$

  The first x coordinate value of **a** subtracted from the first x coordinate value of **b** is smaller than 0.

- Function # 7.

  $\exists \, (a, \ldots p_1, p_n, \ldots b)$, where $d[b, o] > d[a, c]$. For three or more data points, the distance between at least one of the points (intervening between the first and the last data points) and a point **o** on the line joining the first, **a** and the last, **b**, data points is greater than the value of the parameter DIST.

- Function # 8.

  $\exists \, (a, \ldots p_1, p_n, \ldots b)$, where $d[a, b] > LENGTH1$.

  The distance between the data points **a** and **b**, separated by *at least* one other data point, is greater than the value of the parameter LENGTH1.

- Function # 9.

  $\exists \, (a, \ldots p_1, p_n, \ldots b, \ldots p_2, p_n, \ldots c) \ni \bigcirc$ (with radius, RADIUS1)

  The three data points **a**, **b** and **c**, where **a** and **b** and **b** and **c** are each separated by *at least* one other data point, cannot be contained within a circle whose radius is given by the value of the parameter RADIUS1.

- Function # 10.

  $\exists \, (a, \ldots p_1, p_n, \ldots b, \ldots p_2, p_n, \ldots c)$ where $\angle abc < \pi - \epsilon$ or $\angle abc > \pi + \epsilon$. The three data points **a**, **b** and **c**, where **a** and **b**, and **b** and **c** are both separated by *at least one* other data point, form an angle such that the value of this angle is less than the value of the expression $(\pi - \epsilon)$ and greater than the value of the expression $(\pi + \epsilon)$.

- Function # 11.

  $\exists \, (a, \ldots p_1, p_n, \ldots b, \ldots p_2, p_n, \ldots c)$, where $\triangle > AREA1$

  The three data points **a**, **b** and **c**, where **a** and **b** and **b** and **c** are each separated by *at least* one other data point, constitute the vertices of a triangle with an area greater than the value of the parameter AREA1.

- Function # 12.

  $\exists \, ((x_i, y_i) = \mathbf{a}), \ldots p_n, \ldots ((x_j, y_j) = \mathbf{b})$, where $x_j - x_i < 0$

  Given that the data points **a** and **b** are separated by *at least one* other data point, the first coordinate value of **a** subtracted from the first coordinate value of **b** should yield a result less than 0.

- Function # 13.

  $\exists \, (a, \ldots p_1, p_n, \ldots b)$, where

  (i) $d[a, \ldots p_1, p_n, \ldots b] > LENGTH1$ *and*

  (ii) $d[c, \; p_1, p_n, \ldots d] < LENGTH2$

228

This function is a conjunction, and both parts of the function have to yield true before the function as a whole yields true. First, the two data points a and b, separated by *at least one* other data point, are a distance greater than the value of the parameter LENGTH1 apart. Second, the two data point c and d (which might or might not be the same two data points as a and b) are a distance smaller than the value of the parameter LENGTH2 apart.

- Function # 14.
  $\exists (a,...p_1,p_n,...b,...p_2,p_n, ...c)$ where
  (i) $(a,... p_1,p_n ...b,... p_2,p_n ...c) \ni \bigcirc$ (with radius, RADIUS1) *and*
  (ii) $(d,...p_1,p_n,...e,...p_2,p_n, ...f) \in \bigcirc$ (with radius, RADIUS2).
  This function is a conjunction, and both parts of the function have to yield true before the function as a whole yields true. First, the three data points, a, b and c, where a and b, and b and c are both separated by *at least* one other data point, *cannot be contained* within a circle whose radius is given by the value of the parameter RADIUS1. Second, the three data points d, e and f (which might or might not be the same as the three data points a, b and c), *can be contained* with in a circle whose radius is given by the value of the parameter RADIUS2.

- Function # 15.
  $\exists (a,... p_1,p_n ...b,... p_2,p_n ...c)$ where
  (i), $(a,... p_1,p_n ... b,... p_2,p_n ... c) \rightarrow \triangle$, area $>$ AREA1, *and*
  (ii), $(d,... p_1,p_n ... e,... p_2,p_n ... f) \rightarrow \triangle$, area $<$ AREA2
  This function is a conjunction, and both parts of the function have to yield true before the function as a whole yields true. First, the three data points, a, b and c, where a and b, and b and c are both separated by *at least* one other data point, are the vertices of a triangle whose area ia greater than the value of the parameter AREA1. Second, the three data points d, e and f (which might or might not be the same as the three data points a, b and c), are the vertices of a triangle with an area less than the value of the parameter AREA2.

3.2 An implementation example

We are currently investigating network implementations of these 15 functions. In one part of our research we are exploring ways to implement each of the functions as a reliable network. At time of writing, we have organized these 15 functions into 'similarity' sets with respect to network implementations, and are exploring the ease and reliability of implementing each as a neural network. Given some minimal preprocessing (mostly the splitting of a sequence of consecutive data points into a set of all pairs), we have found it relatively quick and easy to generate network implementations of nearly all 15 functions. But function # 4, for example, is proving difficult to implement with more than about 70% reliability on individual networks.

In the other part of our current investigation we are exploring multiversion reliability and methodological diversity using implementations of these component functions, in particular Function # 1. The fact that a neural net can be readily trained to accurately implement this function is, at first sight, counterintuitive and thus quite surprising. For, although neural nets are known to be effective with pattern-recognition tasks and Function # 1 looks like such a task, the training input is simply 5 real values – 4 for the two xy coordinate points, and 1 for the value of the LENGTH1 parameter. The network is given *no information about how to interpret these values*, except that some groups of 5 values are paired with the output TRUE, and others are paired with the output FALSE. Nevertheless, a 5-7-1 network (i.e. 5 input units, 7 hidden units, and 1 output unit) quickly learns a training set (1000 randomly generated, training pairs takes from 4000 to 14000 iterations, or 10 to 30 minutes on a Silicon Graphics, IRIS Indigo), and is around 95% correct when tested on 500 randomly generated, previously unseen, inputs.

In order to begin our investigation of multiversion software reliability with this function, we devised 4 'methodologies' for network programming. A set of 10 nets was trained with each of the 4 methodologies.

set f: fixed training set of 1000 randomly generated; fixed weight initialisation; vary architecture (number of output units varied from 1 to 10)

set ts: varied training, 1000 randomly generated, nonoverlapping for each network; fixed weight initialisation; fixed architecture

set ws: fixed training set of 1000 randomly generated; varied weight initialisation, different random seed for each network; fixed architecture

set rt: as for set ts, except that training sets contained only 100 patterns and were constructed 'rationally' rather than purely randomly (i.e. systematic coverage of the problem space with both barely legal and barely illegal training pairs)

As a result we had 40 individual networks, trained to compute Function #1. They were divided into 4 sets, each containing 10 versions developed using a particular methodology. The initial questions for which we are seeking answers are: How diverse are these methodologies? What can we learn about maximising such diversity within the neural computing paradigm? How does the diversity between sets compare with the diversity within sets? What can we learn about increasing software system reliability with the multiversion approach in neural computing?

Some preliminary answers are emerging.

4 First results

The statistical results are computed using the model of Littlewood and Miller ([3]).

The reliability of each of the 40 trained networks was assessed by testing each of them on 500 randomly generated, previously unseen test inputs. The reliability, using this strategy, was found to vary from 91.0% (for one of the rt versions) to 94.6% (for one of the ts versions). A first surprise was that the rt set, which was expected to be the 'best' implementation of the function (because each individual version had learned selected boundary cases), came out worst in the random testing – the average performance of the individual versions was the lowest of the four sets (it also contained the individually worst version, mentioned above). However, there was less than 4% variation in the tested performance over the entire 40 nets.

For comparison purposes, an identical network was trained on all 10000 training examples (the same 10000 that are spread over ten networks in set ts). It was then tested on the same 500 test inputs, and proved to be 95.8% correct. This is significantly better than any single one of the trained nets in the ts set, but, of course, training took a lot longer, and having only one net excludes the exploitation of system reliability increases through a multiversion approach. For example, the probability that just two, randomly selected versions from the ts set are correct is 96.4% (see below). This result suggests that, for a given training set, it is a more effective strategy (with respect to overall system reliability) to train a set of networks on subsets of the total training set and to operate a multiversion strategy, than to train one network with the total training set.

The general question that underlies the issues of training set size versus coverage of the function by the trained net is: How does training set size effect 'generalization' (i.e. the percentage of random test cases computed correctly after training)? In order to examine this question a net was continually trained, tested, and retrained on successively larger training sets, starting with 50 training patterns and working up in steps of 100 (after first step of 50) to finish with 1000 training patterns. As expected, more training tends to produce better coverage of the function. In addition, the biggest gains are made early on, and then it becomes progressively more difficult to increase coverage by increasing training set size. Figure 1 illustrates tha exact nature of this behaviour. It is not clear that the early maximum in generalization (at about 200 training patterns) is of general significance. Like so much of this data, it is population averages that are indicative; any single net may exhibit some idiosyncracies.

In the process of training networks to learn all of the training patterns, a similar law of diminishing returns is usually observed – i.e. the bulk of the training set is learned quickly, but it becomes successively more time consuming to learn each extra few en route to learning all of the training examples. An obvious question is: What is the trade-off between percentage of training set learnt and the reliability of the partially trained network. Our data on this (again using Function # 1) shows, for example, that after 100 iterations of the training data, 651 training patterns (out of a total of 1000) had been learnt, and the network when tested on 500 randomly generated test patterns, was 62.6% correct. After 1000 iterations, 905 patterns had been learnt, and the same test gave 89.6% correct. After 5000 iterations, 983 patterns had been learnt, and when tested the net was 94.2% correct. Finally, after just over 10000 iterations, all of the 1000 training patterns was learnt, and the resultant net was 95.6% correct. In this study 95% of the final reliability was obtained with only 14% of training effort (measured as iterations of the learning cycle) and having properly learnt only 93.5% of the training set. This suggests that, provided the extra reliability can be gained through multiversions, learning something less than the total training set may be an effective practical strategy for use with individual network versions.

In agreement with all earlier studies of multiversion software reliability, we found evidence of many common errors within the sets of alternative versions – e.g. probability of two randomly selected versions (from within any one set, x) failing on a randomly selected input, $E(\theta_x^2)$, is much greater than square of the probability that one randomly selected version will fail on a randomly selected input, $E(\theta_x)^2$. Representative values are:

$$E(\theta_{ts}) \quad = \quad 0.0708$$
$$E(\theta_{ts})^2 \quad = \quad 0.00501$$
$$E(\theta_{ts}^2) \quad = \quad 0.0356$$

However, the best set, in this respect (i.e. the most diverse set), was the rt set. For, although single version failure probability was the *highest* of any of the 4 sets, $E(\theta_{rt}) = 0.0796$, the probability of two-version failure, $E(\theta_{rt}^2) = 0.0315$, was the *lowest* in any the 4 sets! This indicates a lack of common failures between individual versions, and how a simple multiversion strategy, applied to a set of diverse versions, more than compensates for individually weak implementations.

As was anticipated, the ws and f sets displayed little diversity, and added correspondingly little reliability when used in conjunction with each other, or the ts set, in a cross-methodology approach to multiversion reliability. One of the suggestions of Littlewood and Miller was that the undermining of multiversion reliability through the occurrence of common errors could be avoided by selecting individual versions from methodologically diverse sets of alternatives. Progress with this approach can be assessed by computing the probability of two randomly selected versions (one from each set, x and y) failing on a randomly chosen input, $E(\theta_x, \theta_y)$.

Computations of this quantity between pairs of sets gave:

$$E(\theta_{ts}, \theta_f) \quad = \quad 0.0368$$
$$E(\theta_{ws}, \theta_f) \quad = \quad 0.0535$$
$$E(\theta_{ws}, \theta_{ts}) \quad = \quad 0.0376$$
$$E(\theta_{rt}, \theta_f) \quad = \quad 0.0065$$
$$E(\theta_{rt}, \theta_{ts}) \quad = \quad 0.0069$$
$$E(\theta_{rt}, \theta_{ws}) \quad = \quad 0.0068$$

which should be compared with the product of single version failure in each set:

$$E(\theta_{ts}) * E(\theta_f) \quad = \quad 0.00515$$
$$E(\theta_{ws}) * E(\theta_f) \quad = \quad 0.00527$$
$$E(\theta_{ws}) * E(\theta_{ts}) \quad = \quad 0.00512$$
$$E(\theta_{rt}) * E(\theta_f) \quad = \quad 0.00579$$
$$E(\theta_{rt}) * E(\theta_{ts}) \quad = \quad 0.00564$$
$$E(\theta_{rt}) * E(\theta_{ws}) \quad = \quad 0.00576$$

These results are indicative of a lack of methodological diversity between the pairs of sets which do not involve the rt set. This is because the actual probability of failure of two versions, selected one from each set, is an order of magnitude higher than the simple product computation which is based on an assumption of independent failure within each set. But, quite strikingly, once the rt set is involved, we see the actual probabilities of two-version, between-set failure dropping to values quite close to those predicted by the 'ideal' case of totally independent failures.

Computation of the covariance and correlation coefficients of common failure between set x and y, $Cov(\theta_x, \theta_y)$ and $\rho(\theta_x, \theta_y)$ respectively, indicates that the most diversity occurs between the rt and f set, which is a little surprising. The computed values are:

$$Cov(\theta_{rt}, \theta_f) = 0.000705$$
$$\rho(\theta_{rt}, \theta_f) = 0.0199$$

This result is, perhaps, surprising because it is the f set involved, rather than the ts set. The only source of diversity in the f set is through a minor variation in the network architecture (successive repetitions of the output unit), whereas the ts set can exploit the diversity to be found in 10 non-overlapping, randomly generated training sets. However, the point is, perhaps, that this diversity measure is between sets, and it so happens that the f-set generation strategy causes rather different errors to the rt-set strategy, whereas the ts set (although in itself diverse) tends to duplicate the errors in the rt set. The data bears this out: 64 test inputs (out of 1000) failed on one or more versions in *both* the rt and ts sets, but only 38 failed similarly within the rt and f sets.

Moving to other possible strategies: we can consider the probability of a majority of versions being correct. Within the ts set, the sixth moment, $E(\theta_{ts}^6)$, the probability that 6 randomly selected versions will all be incorrect is 0.0069, which seems to indicate that there is a 99.3% probability that a majority vote (in 10 versions) will be correct. While the between-sets strategy gives that probability of 12 randomly selected versions failing (6 from rt set, and 6 from ts set), $E(\theta_{rt}^6, \theta_{ts}^6) = 0.00002$, which suggests that a majority vote across these two sets will have a 99.998% probability of being correct.

A more demanding strategy might be to select a single one that is correct. In this case, the relevant data, the probability that all 10 versions are incorrect in the rt set is 0.0005. And the probability that all 20 versions, in both the rt and the ts sets, are incorrect is $0.47 * 10^{-6}$. This last figure suggests that, if we could select the right version from within the two sets, then we would have a very reliable system. We plan to look at ways to train a meta-net as a 'switch' to select correct versions.

5 Conclusions and discussion

At the moment, in this very early stage of the research we are unable to draw many firm conclusions. For the most part, we have only just begun to explore the possibilities. So little is known about the scope, limitations, and capabilities for training neural nets to implement well-defined functions, almost all experiments are unique. And thus all results have to be thoroughly studied, and repeated using a variety of nets and functions, before it will be possible to separate the various idiosyncrasies of individual nets from the general characteristics of this technology as a whole.

In addition, there is a need, which we acknowledge and are addressing, to develop a proper understanding of how and why the observed phenomena occur. Such an understanding will lead to a predictive science, a proper basis for a practical technology of neural computing. We need, for example, to be

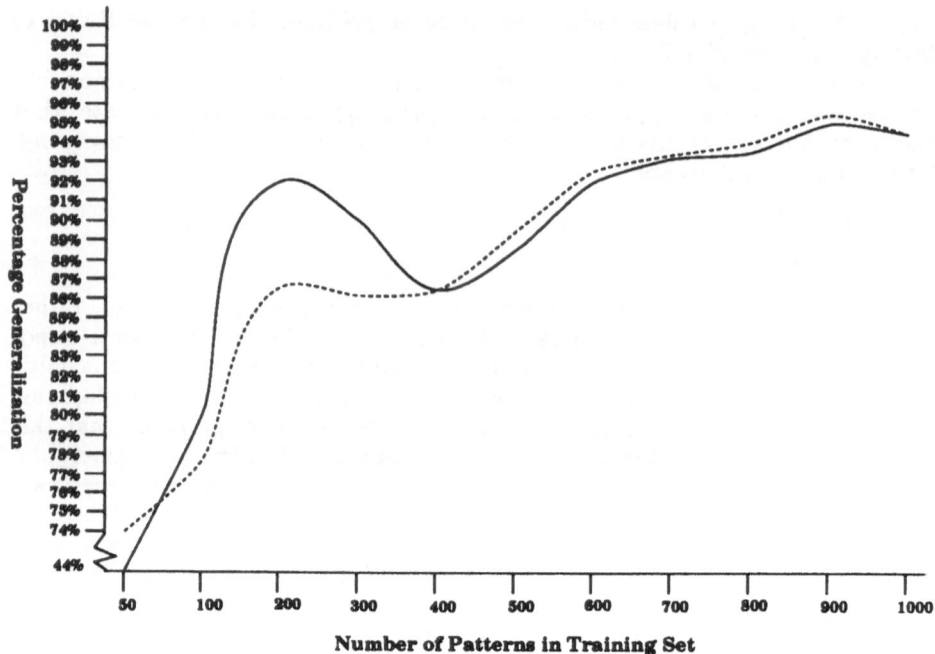

Figure 1: Graphical display of the results of the incremental training experiment. The thick lines represent the generalization performance of the net when tested on test sets half the size of the training set; the dashed line indicates generalization performance of the net when tested on a single, 500 pattern test set.

able to 'examine' a trained network and determine precisely what function the training has implemented, rather than just rely on testing to tell us this. We are actively exploring the use of visualization techniques to facilitate a proper understanding of neural computing [9].

Quite apart from an elucidation of a technology for 'network programming', the pursuit of such a radically different implementation technology has cast some interesting light on the nature of conventional software engineering itself [5], [4]. For instance, the role of the specification: the keystone for the entire process? a misinterpretation of the real requirements? etc. [6] Rather than build an implementation to satisfy a specification, we build an implementation from requirements (represented as a set of input-output pairs) and abstract the specification of what's been implemented.

Similarly, the radical shift in the economics (and practicalities) of developing multiple versions of an implementation opens totally new pespectives on software reliability. For example, distributing the development effort over a population of individually, insufficiently reliable versions may be a viable alternative to concentrating all effort on the production of one sufficiently reliable version. This reasearch will shed light on these sorts of general issues as well as on the more specific ones of methodological diversity within, and between, programming paradigms.

References

[1] J. M. Adams and A. Taha, "An Experiment in Software Redundancy with Diverse Methodologies," *Proc. of the Twenty-Fifth Hawaii International Conf. on Systems Sciences*, 1992.

[2] J. C. Knight and N. G. Leveson, "An Experimental Evaluation of Independence in Multiversion Programming," *Trans. on Software Eng.*, vol. SE-12, no. 1, January, 1986.

[3] B. Littlewood and D. R. Miller, "Conceptual modeling of coincident failures in multiversion software," *IEEE Trans. on Software Engineering*, vol. 15, no. 12, December, 1989.

[4] D. Partridge, *Engineering Artificial Intelligence Software*, Intellect Books, Oxford, UK, 1992.

[5] D. Partridge, "On the difficulty of really considering radical novelties," Technical report, no. 238, Dept. Computer Science, University of Exeter, EX4 4PT, UK, 1992.

[6] D. Partridge and A. Galton, "The specification of 'specification'," Technical Report, no. 262, Dept. Computer Science, University of Exeter, EX4 4PT, UK, 1993.

[7] D. Partridge and N. E. Sharkey, "Neural computing for software reliability," *Expert Systems Journal*, [submitted].

[8] D. E. Rumelhart, G. E. Hinton and R. J. Williams, "Learning internal representations by error propagation," In D.E. Rumelhart and J.L. McClelland (Eds), *Parallel Distributed Processing*, Vol. 1, MIT Press, Cambridge, MA, 1986.

[9] N. E. Sharkey, "Connectionist representation techniques," *AI Review*, vol. 5, pp. 143-167, 1991.

Knowledge Based Systems in Safety Critical Applications

Audrey Canning
ERA Technology Ltd
Leatherhead, Surrey, England

Abstract

Recent work at ERA Technology has shown that Artificial Intelligence is being used to develop systems which are likely to have safety implications. Far from being unacceptable, our investigations show that, if properly developed, many of these systems have characteristics desirable in conventional safety systems. Moreover, since the approach tends to focus on the problem definition stage of a project, it is possible that the techniques could provide a valuable insight into requirements capture and system validation. In this paper the reasons for our interest in Artificial Intelligence are explained, and a comparison between their capabilities and more conventional approaches is made with respect to desirable safety characteristics. We conclude by presenting an overview of a collaborative research project intended to further our understanding of the issues.

1. Introduction

Increasing attention is being given to ensuring high levels of trustworthiness in computer systems used in situations where their malfunction can cause loss of life and/or severe damage to the environment. Although techniques are available which can formally prove that a software program meets its specification, such techniques are expensive to use and do not deal with the ability of the specification to meet the requirements of the application. Nor indeed can such techniques address the validation of the requirements model itself, which has long been recognised as a significant source of error in computer system developments [1,2]. There is a need therefore to continue to seek development paradigms which extend the coverage of software integrity issues, even to the definition of the behavioural models from which the requirements for the control, monitoring and safety systems are derived.

Over the last ten years a number of techniques, classified under the general subject of Artificial Intelligence (AI), have become of interest for the rapid development of control and advisory systems. This is particularly the case where the systems are intended for use in new areas of application, or where the problem model is not well understood. In the early 1980s, considerable interest was directed at the development of knowledge based systems (KBS) as a means to capture the processes carried out by the human experts within a complex control or monitoring problem. Whilst the publicity associated with this type of system has decreased, a recent study carried out within ERA Technology has confirmed that KBS are under development in a wide range of industries. Notably, for safety purposes, such domains include flight control, railway interlocking, power generation and monitoring, and robotic control.

More recently there has been widespread industrial interest in the use of neural networks and fuzzy logic for control applications. Neural networks in particular, are perceived as potentially having a huge impact on engineering practices. In one trial application, for example, sonar target characteristics apparently were identified in three hours using a neural network algorithm, compared with ten months to derive the information from human expertise [3]. The main advantages of the neural network are considered to be the parallel nature of the architecture, which facilitates considerable increases in processing capability, and the generalised learning capability. The latter feature in particular could radically alter the traditional engineering approach to identifying complex engineering requirements, namely through tailoring of scientific models, backed up with human expertise, simulation and modelling.

During the work at ERA on advanced software systems for safety critical applications, we have had cause to note a similarity between certain types of industrial safety systems and some AI systems, notably KBS. We have also encountered significant difficulty in the validation of large data base systems using conventional software testing techniques, particularly in achieving adequate test coverage for all data input conditions. For these reasons we have become interested in techniques used to verify and validate KBS. Further, since the primary focus during the development of AI type systems is on a better understanding of the problem domain, we have become interested in the effect the use of this type of system would have on the safety of the overall system. Indeed, our investigations, far from confirming the view that AI systems cannot be considered for critical applications, have led us to the view that there are several features of an AI approach which could have a beneficial effect on safety. In the remainder of this paper the reason for our interest in AI systems is explained, and a comparison between their capabilities and more conventional approaches is made with respect to desirable safety characteristics.

2. Requirements for Safety

Generally by the very nature of their inductive development process, systems developed using AI techniques tend to show excellent performance against the particular data model used during the problem definition activity. Moreover, this performance can often be achieved for the expenditure of very low levels of effort when compared with conventional software systems [4]. For example, some estimates have placed the cost of development of AI type systems at perhaps one tenth to one hundredth of that associated with conventional systems to achieve the same purpose, moreover the maintenance effort associated with the deployment of AI technology can also be very low, indicating a considerable level of user satisfaction with such systems once deployed. Where there are significant safety risks associated with the deployment of computer systems however, it is necessary to justify that the proving process has been sufficiently broad ranging as to encompass all credible hazardous conditions likely to be encountered during the operation of the system.

For software systems, the level of design complexity is so great that demonstration against particular instances of the operational data conditions is not generally considered an adequate demonstration that the system will behave safely under all likely operating conditions. Instead, most existing safety standardisation initiatives require that the processes used during the software development activity can be shown to adhere to 'good practice'. These initiatives are founded on the belief that adherence to 'good practice' will reduce the numbers of errors within a piece of software. In the field of software engineering the identification of 'good practice' is still evolving, however there are certain general criteria which need to be shown to justify deployment of any safety critical system including [5]:

-   that adequate information concerning the system is available to achieve safe use

-   that a valid version of the system is in use

-   that the equipment is fit-for-purpose

-   that all reasonable steps have been taken to ensure safe use

3.   Expert Systems and Safety

3.1. Performance Against Good Safety Practice

At ERA we have carried out a comparison of the extent to which the above requirements are likely to be satisfied by software systems developed using conventional methods and those developed using AI techniques. Generally, it is arguable that the first criterion above is unlikely to favour systems developed using a conventional lifecycle methodology over those based on AI techniques, since the AI approach tends to ensure a much closer interaction with the personnel experienced in the end use of the proposed system. For example, the greater emphasis placed on understanding the problem definition during the AI development process, is likely to result in a better understanding of the proposed use of the system by the designer, which in turn places him in a position to provide more comprehensive information regarding the safe use of the system to the end user. By contrast, our experience during independent safety proving of conventional systems suggests that identification of the operating procedures is often something of an afterthought.

With regard to use of the correct version of the software, this is an area where considerable effort has been directed at standardisation in an effort to make the results of software modification more visible to outside review, and to control the difficulties associated with the development of large and complex logical designs. Clearly any system which allows evolution and modification of the software product

238

from that approved during development of the safety case will give rise to concern. For this reason the use of an AI system which incorporates mechanisms for learning and self modification after the system has been approved will be very difficult to justify within a safety application. However, in many AI development programmes learning algorithms are used only during the development process, and therefore the problem is commensurate with that experienced for conventional software. However, there is a danger that where such mechanisms can be readily incorporated into the deliverable system, a practicioner not familiar with safety requirements could inadvertently design in self modification facilities. Hence, guide-lines are required to address the identification and deployment of any type of system incorporating learning algorithms, which should apply equally to conventional and to AI development practices. Indeed, the basic guidance might also apply to manual modifications. For example there is often a strong operational need to be able to modify the values of parameters used during control of the system. Such modifications are often justified by allowing only limited deviation within a defined safe operational boundary, which practice could equally apply to automatic learning algorithms. The means of identification of such a boundary is not trivial however, since digital systems are likely to exhibit discontinuous behaviour.

Assessment of fitness-for-purpose is achieved through systematic examination of the behaviour of a system. Testing, inspection, and analysis techniques may all be required to demonstrate adequate levels of fitness-for-purpose. For conventional software the prevailing opinion is that use of recognised design practices, control and information flow analysis, and dynamic testing needs to be satisfactorily undertaken to show fitness-for-purpose. There is also support in some industries for the use of semantic analysis against a (de-facto) formal specification.

The wide acceptance of the testing practices listed above is such that they have now been adopted by most software standardisation initiatives. As a result these practices, arising directly from the practices deemed feasible and appropriate for development of conventional software, will also be used to assess the fitness-for-purpose of systems developed using AI techniques. This approach can only result in heavy bias against an AI system, since verification techniques are of limited use where, by definition, there is no 'a priori' specification (gold standard) against which verification can be carried out. For example, work within the SEMSPLC project [6] has shown that the use of control flow analysis is of limited application for certain types of data-driven system. Nevertheless, a wide range of verification procedures have been proposed for AI systems [eg. 7], notable amongst these being the analysis of the internal states of the rule based system [8]. Moreover, the differing background and experience of the staff involved in AI development work has resulted in an emphasis on assessment techniques which compare a system performance with that achieved by human activity for the same task. Consequently, most AI systems are assessed against a subjective perception of usability and usefulness. Since such techniques are more likely to identify errors made during the requirements capture stage of a project than conventional verification, arguably, for system validation, AI practices may well be more onerous than conventional software development standards. This argument is particularly strong in areas where the problem is not well defined, and indeed would apply to all development processes where ad-hoc animation, simulation and modelling are currently the only approaches to validation.

3.2 Opportunities for Improved Safety

In view of the preceding discussion, the fourth criterion above, to show that all reasonable steps have been taken to ensure safety, would appear to have the effect that the use of AI systems can only be justified in non-critical applications. In particular, whilst judgement of the extent to which a system meets this criterion needs to be determined through expert debate, the very strong bias towards conventional practices and lifecycles in current standards will make it difficult to gain credit for any other approach, even where other advantages for safety can be demonstrated.

To accept this position will, in our view, be to the long term detriment of the UK software engineering community. Apart from the commercial advantages of reduced development time, AI techniques offer many opportunities for improved integrity within the overall system, including:

- the use of symbolic, declarative and application oriented languages

- the means to explore safe operational boundaries

- separation of application knowledge from its interpretation by the language kernel

- explicit use of predicate logic as the basis of the decision making algorithms

- the use of algorithms derived from relevant problem data rather than general scientific models

- the use of a prototyping approach to development.

It is our contention that the use of declarative and symbolic languages allows the operator a better opportunity to develop a useful mental model of the behaviour of the control system and of it's interaction with the plant. Rules expressed in English can be readily communicated across engineering disciplines and skill levels, whilst retaining their logical accuracy. This in turn increases the opportunity for a common understanding of the system and its behaviour. The use of explicit inference algorithms in conjunction with symbolic languages is such that the individual steps within a computational algorithm can be listed to provide a form of explanation as to how the results of the computing process were achieved. This in turn provides better support for the operator when attempting to confirm, under pressure, the actions taken by the automatic system. Clearly, with human error being a major contributory factor to safety incidents [9] this feature alone is potentially of great value in achieving improvements in the overall safety of a system. Moreover, the use of logic based representational techniques is particularly appropriate for many safety systems, where the safety requirement is often expressed in the form of interlock or shut-down logic.

240

Our recent enquiries show that a significant focus of industrial interest in AI techniques for safety critical applications is in their use to explore the boundary of system behaviour and safe operation. In particular, many engineering endeavours involve areas where fundamental scientific and physical models of the environment are still poorly understood. Typical of such applications are global weather systems, nuclear reactions, electromagnetic radiation and robot geometry. Nevertheless, in order to carry out certain necessary engineering endeavours, reasonable models of these processes must be developed and used, often with little opportunity for off-line validation. Traditionally, simulation has been used to validate such application models, yet these are likely to be based on a limited number of runs, which may not cover all the conditions likely to be encountered in a real situation. AI techniques are already being used to capture improved information about such applications, and there is reason to believe that information captured in this way is of benefit in developing a better understanding of the underlying scientific models. Nevertheless, since the information collected could influence the way in which hazardous operations are carried out, there remains a critical need to be able to validate the appropriateness of the application models.

Separation of the application knowledge from its interpretation has the advantage that, to some extent, the application model can be analysed without regard to the effects of limitations of the expressive adequacy (power) of the implementation technology. Indeed, the approach can be considered as a form of abstraction, which is widely used as a means to reduce complexity and improve visibility during the specification of safety critical systems. Conventional examples using separation of application from it's interpretation include the MASCOT run time kernel, programmable logic controllers and SCADA systems. Clearly, the ability of modelling languages to describe behavioural models will have an effect on the information which can be incorporated in the application model, however AI approaches include a wide range of different knowledge representation techniques. Indeed, there are even taxonomies of knowledge [eg 10], which can be used to refine the information content to the appropriate level for the implementation technology.

An added benefit of the separation of application knowledge from it's interpretation is that the interpreter software can be re-used in many applications. Although unproven in scientific trials, it seems reasonable to suppose that wide re-use of the software in different applications is likely to lead to a greater exposure to different input conditions, and therefore more likely to reveal latent errors. With appropriate infrastructure to record and remove errors, the overall integrity of the software may be improved. A further advantage is that often the language kernel software is relatively straightforward and may even be amenable to specification and analysis using formal mathematical logic.

A feature of certain AI systems is the use of predicate logic as the basis of the reasoning algorithms. The mathematical basis for such systems is equivalent to that recommended for formal mathematical specification, although the approach is inductive rather than deductive in nature. Indeed it would appear that the major difference between conventional software engineering and the inductive 'knowledge based' approach is as follows

- in conventional software engineering a theoretical mathematical model of the problem is selected and then the parameters of that model are 'adjusted' until the model fits the observed or postulated system behaviour

- in the inductive approach, the logical model is generated directly from postulated or observed examples of system behaviour to obtain a general solution.

In the first case the boundaries within which the system will behave in the required manner are generally well understood, and the expressive adequacy of the top level specification is selected so as to be capable of defining the important relationships between variables. Nevertheless, once the model is digitised, the expressive capability of the particular set of logical axioms selected will radically affect the relationships which can be described. In the second case, a particular set of logical axioms is adopted from the start, and although the axioms need be no less expressive than the 'top down' approach, the analysis of the bounds within which the algorithm applies may be less apparent. Whichever approach is adopted there is an equivalent need to ensure that the final digital algorithm accurately reflects the real world behavioural relationships.

Certification of AI systems has often not been considered possible due to the absence of generally acceptable lifecycle models. In particular, a prototyping approach to software development is used, which is contrary to established guide-lines. Nevertheless, there is a growing body of opinion which advocates that lifecycles based on prototyping, if properly controlled, can be of considerable assistance in detecting errors early in the development process. In fact, the prototyping approach is entirely compatible with the need to improve the requirements capture stage. Moreover, there are a growing number of methods and techniques [eg 11] for systematic development of AI type systems. Whilst these techniques are aimed at situations where non-engineers are the source of knowledge about the system behaviour, nevertheless, their availability should provide a basis for quality control of AI techniques used in safety critical systems.

Although many of the above arguments do suggest that AI has a role to play in overall system safety, there are many issues which have still to be considered. For example, the use of these techniques may have a radical affect on the human role within a safety system. Certain types of AI system may lead to de-skilling, since the installation of such a system may obviate the need for day-to-day involvement in safety routines. If it then becomes necessary to have such experience, there may be a reduced level of expertise available to deal with an incident. The characteristics of AI languages, and the way in which semantic properties and their exceptions are handled, needs to be explored, together with validation techniques for the language kernel. Moreover, some types of AI system can have a filtering effect on data, which although beneficial in reducing the complexity of the operator task could result in masking of vital information. Introduction of AI type systems will require the careful consideration of the effect on the whole organisational system, of which the automated systems and their implementation techniques will form merely one part.

4.   ISSAFE : Inductive Software Solutions for Use in the Flight Environment

From the forgoing discussion, we have concluded that there is a need to investigate and understand both the positive and negative effects of AI on critical systems. Not only could such systems increase safety margins, but their cost-effectiveness is such that there is a growing number of systems under development which will find uses in critical applications. In order to ensure that the UK remains competitive in an international market, there is a need, not only to introduce effective controls as to the circumstances under which such systems may be deployed, but also to ensure that companies are not unnecessarily discouraged from the development and use of these systems. Finally, since many of the factors affecting the safety of AI systems are also experienced in conventional data base systems, techniques developed within the AI community may well have a role to play in conventional software engineering technology.

To develop these ideas further, a consortium of companies has recently been formed under the joint DTI/SERC Safety Critical Systems Programme to undertake a collaborative research programme. Part of the objectives of the research programme are to develop a methodology for creating KBS specifications which are amenable to system validation and software verification. Currently, in addition to ERA Technology, the partner organisations include BAe Airbus Ltd., DRA Bedford, Glasgow Caledonian University Company Ltd., the University of Hertfordshire and Westland Helicopters Ltd. In addition, a further six major organisations are sponsoring the work programme[1].

The objective of the research programme is to provide methods and guide-lines for the development and use of KBS in safety critical and high integrity applications, and is organised in three phases as follows:

Phase 1

-    investigate the capabilities of inductive techniques for software specifications to characterise real world safety critical systems in comparison to the capabilities of other systems modelling techniques including time and frequency domain models, Markov models, Petri nets and neural networks

-    assess the plausibility of validating KBS by comparison with algorithmic methods

---

1 Sponsoring organisations include British Nuclear Fuels plc, Docklands Light Railway Ltd., Industrial Control Services plc, London Electricity plc, London Underground Ltd., and Rolls Royce and Associates Ltd.

Phase 2

- prepare guide-lines for the evaluation of KBS for use in safety related systems of differing levels of criticality

- assess the guide-lines by application to a number of based and non-based systems

Phase 3

- evaluate the KBS certification procedures for wide applicability

- assess the strengths and weaknesses in using KBS for safety critical applications in relation to national and international standards

By comparing the different representation methods and by understanding the difference between the deductive 'embedded logic' approach in mainstream software engineering and the inductive creation of the logical model from examples of the interaction of the problem parameters of the knowledge based approach, it is intended to make recommendations as to the types of high integrity systems in which KBS techniques could be used, and the areas where AI technology is likely to present limitations.

It is the intention that these recommendations will be brought together in the form of draft guide-lines and evaluated against actual KBS to assess the extent of their applicability. The results of this work should also indicate the strengths and weaknesses of using KBS in safety applications of varying degrees of criticality and could form a significant input to national and international standardisation initiatives.

5. Conclusions

Early work carried out at ERA has highlighted the need to address the use of systems based on AI techniques in safety critical applications. In particular, work has shown that three important questions remain to be answered, namely:

- AI based systems are available for use in critical applications; how can they be recognised and controlled?

- AI systems are cost-effective, but should not be deployed under current legislation; how can the benefits be realised?

- AI techniques appear to offer many advantages for overall system safety; how can conventional software engineering practices be improved to include such advantages?

In view of the need a collaborative work programme has been set up, involving more than ten major organisations. Currently, the work is at an early stage and the eventual conclusions are difficult to predict. Nevertheless, initial indications are that improvements in safety can be expected from the ability to capture and validate application expertise, the closer representation of actual behaviour, a reduction of complexity through separation of application knowledge from inference procedures, enhanced opportunities for software re-use and a better visibility of complex system operation. Indeed, there may even be improved opportunities to move away from a process based certification approach, to one based on product certification and hence more closely related to the actual behaviour of the end product.

In order to utilise these benefits, improvements in technology will be required in the following areas:

- provision of proven standards and codes of practice to control the development and deployment of AI technology

- improvement in verification techniques, in particular there may be a need to identify techniques for proving of data driven technology and worst case timing behaviour

- improvement in validation techniques, to establish ways in which specifications developed from observations can be validated against general scientific models, and for determining when sufficient information is available to define a novel application area

- exploration of methods for determining the safe boundary of operation within an application.

During the work it is expected to focus on the provision of methods by which inductive approaches can be validated, and to determine means of identifying and controlling the 'AI Hacker'. The use of the techniques to explore safe boundaries is seen as particularly pertinent for UK industry, but there is also a need to understand the limitations of the inductive logic models, and their effect on different types of system behavioural model. The overall effect of shifting the safety focus from system implementation to operational behaviour will also need to be considered.

References

1. Boehm B.W.
   Software Engineering Economics
   Prentice-Hall, Englewood Cliffs New York, USA 1981, ISBN 013 822 1227, 768pp

2. Jones D.R., Murthy D., Blanchard J.
   Quality and reliability assessment of hardware and software during the total product lifecycle
   Quality and Reliability Engineering International, Vol 8, No.5, pp477-483, Sept-Oct 1992, UK

3. Defence Advanced Research Projects Agency (DARPA)
    Neural networks study Oct 1987 - Feb 1988
    AFCEA International Press, USA, 1988, ISBN 0916159175, 629PP

4. Applications of expert systems, Volume 2
    Based on the proceedings of the third and fourth Australian conferences
    Edited by J Ross Quinlan
    Turing Press Institute in association with Addison-Wesley Publishing Company

5. The Institution of Electrical Engineers
    Safety related systems : Professional brief
    The Public Affairs Board, Institute of Electrical Engineers, Issue No. 2, Sept
    1992.

6. Hedley D., Kirsopp R.G.
    The testing of ladder logic programs for programmable logic controllers
    To be presented at the 1st European International Conference on Software
    testing, Analysis and Review hosted by the British Computer Society
    October 1993

7. Institution of Electrical Engineers
    IEE Colloquium - testing expert systems
    London 1987
    IEE Digest Number 87/110

8. Bench-Capon T., Coenen F., Nwana H., Paton R., Shave M.
    Two aspects of the validation and verification of knowledge based systems
    IEEE Expert Vol 8, No 3, Jun 1993, pp76-81

9. Brazendale J., HSE
    Human error in risk assessment
    HSE/SRD/AEA Technology Feb 1990, 68pp

10. KEMRAS : A knowledge elicitation manual for RAs
    Part II Section 6
    Alvey Project IKBS 098
    ERA Technology Ltd 1989

11. Hichman F.R., Killin J.L. Et Al
    Analysis for knowledge-based systems. A practical guide to the KADS
    methodology
    Wiley and Sons 1989, 190pp

The rôle of Formal Methods in the Engineering of Safety Critical Systems

Dr P J Byers
Reference Information Systems
47 Sycamore Drive
Ash Vale, Aldershot
Hants GU12 5JY

Introduction

The aim of this paper is to present the conclusions of a series of projects in which advanced formal techniques have been applied to assist in the verification of the design of safety-critical systems. One project in particular is discussed in detail. The focus is on the rôle of a formal approach within the safety engineering process and the requirements this places on the formal techniques themselves.

The contribution that formal methods will eventually make to software engineering is still a matter of considerable debate. Whilst advances are still being made in reinforcing the theoretical foundation of many aspects of software engineering, the goal of "utility" formal techniques, as familiar to the software engineer as the mathematics of structures are to the mechanical engineer, is still a long way off. This paper only examines the requirements of small scale, high-integrity systems; it does not address the needs of large scale development projects, for which the rôle and applicability of formal methods are even less well understood.

The first section of the paper describes the context in which techniques have been applied, and the second main section identifies some wider conclusions from this application.

Project experiences

One example project, and some case-specific conclusions, are discussed. The elements of the project discussed are identification of the safety requirement, generation of a system design and production of the formal safety proof.

The requirement

An existing control system uses a high-bandwidth communications link to exchange critical control information between master control processors and slave functional and sensor units. A roll-call protocol is used wherein each slave unit is polled in sequence to issue command messages and responds with status messages. The cyclic nature of the protocol ensures that messages pass in each direction at a constant rate and any failures, either in end-systems or in the communications link, are detected within a bounded time and safety can thus be maintained. The system architecture is shown in figure 1 below.

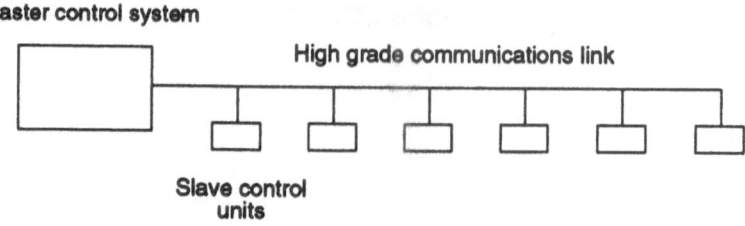

Figure 1: Control system architecture

A new communications system was proposed so that low-grade communications links could be used in place of the existing high grade link. The low grade link could not support the bandwidth required by the existing protocol but could in principle support the level of actual information exchange involved (since most command and status messages would in practice be repetitions of the previous messages and would only be necessary to confirm the continuing correct operation of end-systems and communications bearers).

The new communications subsystem would report changes to command and status messages only across the low grade link in order to minimise the bandwidth required. The design problem was therefore to develop protocols for operation across the low grade link that did not compromise the safety of the whole. The architecture of the proposed system is shown in figure 2.

The safety requirement

The first step in the development process involved identifying the safety requirement for the new communications subsystem. No safety requirement for the communications element of the existing system had explicitly been identified. The aim was to identify the safety requirement *independently* of the functional requirement. (The distinction here is that the functional requirement identifies what the system must do under normal operational conditions, whereas the safety requirement identifies what the system must do *in any event* in order to be safe.)

The safety requirement for the system was formally stated independently of any outline design, based on its behaviour in the context of system interfaces. Formulation of the safety requirement was only possible once its impact on wider system behaviour was analysed in the context of the environment.

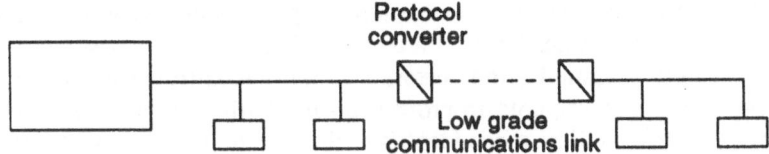

Figure 2: Architecture of proposed system

One example of the influence of external agents concerned the possible requirement for correct sequencing of update messages across the low grade link and the timeliness properties needed. In the existing system, changes in status of remote sensors can actually be reported to the master unit in an order different from that in which they occurred; this is because the ordering of response messages is fixed and independent of the sequence of events in the system's environment. The impact of this on the system is that the details of each installation must be considered to determine the possible effect of incorrectly sequenced status updates and, in practice, the physical separation between sensor units must be considered in conjunction with the timescales over which process in the environment evolve. In this way, detailed aspects of protocol timing are constrained by the behaviour of the environment. When the low grade link is interposed, the intricacies of timing must be considered very carefully in order to define the safety requirement.

The safety requirement was stated in terms of the acceptable forms of behaviour of the system and it related the sequence of messages issued and received at each node. This is therefore a constraint about events distributed in both space and time.

The formal safety requirement was validated using a number of techniques. One example was the analysis of possible failure modes of components of the existing system and using these to define 'safe' modes of failure behaviour of the new communications subsystem. Other techniques involved the development of models of parts of the existing system in order to understand its behaviour more fully.

Design

The design produced involved the specification of two parallel layered protocols, one operating over the high grade link, the other operating over the low grade link. Even though a specification for the high grade link protocol already existed, it needed to be strengthened in order to ensure that various end-to-end synchronisation properties of the system were maintained. The different protocol layers reflected the different timescales over which processes were operating; eg bit-level operations, message-level, the level of reliably sequenced exchanges and full protocol cycles. Interactions between the two protocols also occurred on many timescales; this necessitated elements of the application to operate at each level in the stack. The logical structure of the completed design is shown in figure 3.

Once an outline design was produced it became possible to assess the potential impact of various failure modes of hardware components and each of the possible modes was considered in the design. Hardware and software measures were needed to defend against communications errors and fail-uncontrolled node errors, but mechanisms were also required to protect against initialisation errors, and other latent errors. For each failure mode identified, the time to detection of the particular error was calculated in order to estimate the level of safety achieved. For each particular type of error, the time to detection was proved to fall below a particular threshold.

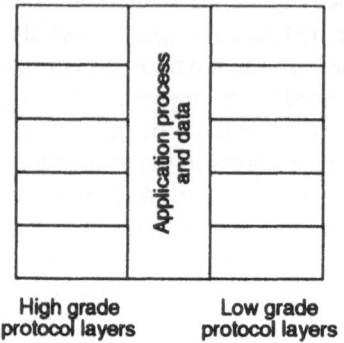

Figure 3: Logical design of communications subsystem

The design (and proof, discussed below) addressed protocol and system behaviour but did not require a full software or hardware design to be produced. Assertions made about software operation execution time, interference and concurrency would then need to be verified during as part of software design. Reliable hardware and software mechanisms to implement monitors, for example, would need to be provided and assured as part of software design.

Proof

Production of the proof consumed about 40% of the total project effort. Even though the design was thoroughly reviewed prior to production of the proof, numerous errors and inconsistencies were identified during the proof process. Whilst most of these would have been detected during software implementation and test, there were some errors that were subtle enough to have escaped detection for some considerable time.

The structure of the argument was intended to be as natural as possible to reflect the intuition behind the design. Two proofs were produced, one of conformance to functional requirements and the other, in much greater detail, of conformance to safety requirements. Both had similar structures, but widely different techniques were used. (For example, safety proofs often involved reasoning 'backwards' to determine the historical cause of particular states, whereas the functional proof more often involves reasoning 'forwards'.) The structure of the two proofs is show in figure 4 below.

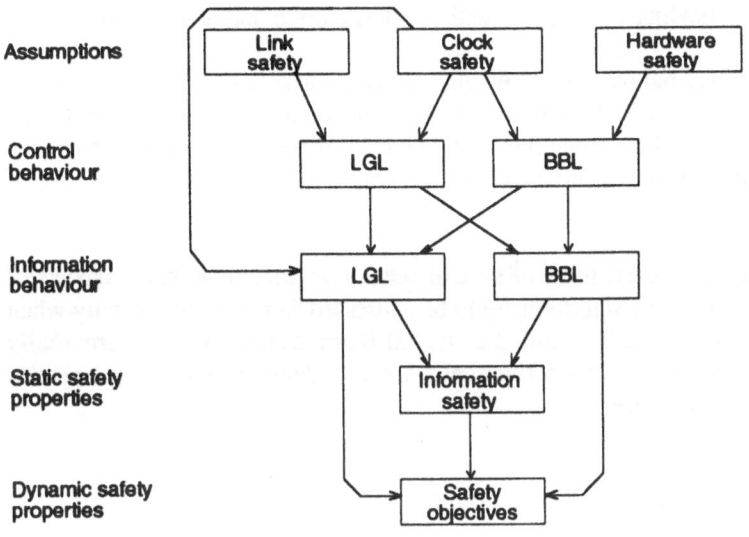

Figure 4a: Structure of safety proof

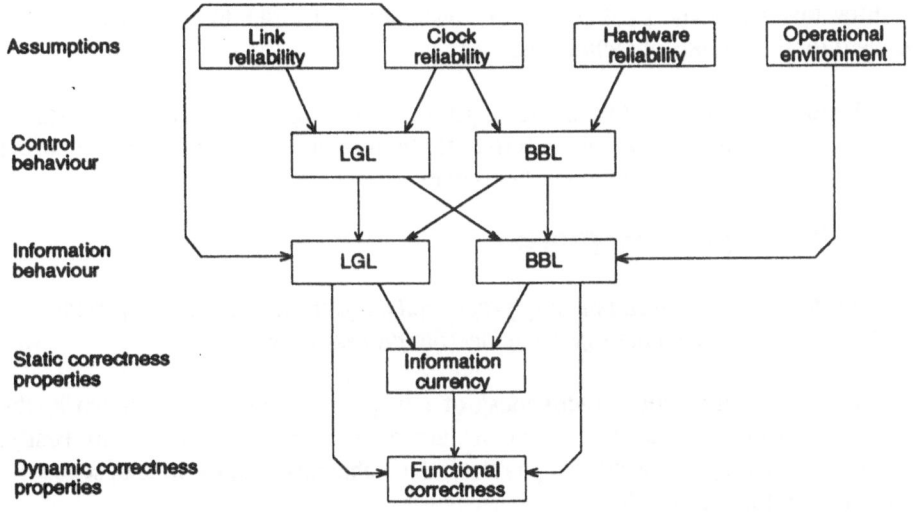

Figure 4b: Structure of functional proof

Each box in the above diagrams corresponds to a collection of related assertions that figure in the proof. The proof relies on particular assumptions about hardware behaviour. These assertions are documented, but cannot be proved; in practice, it will only be possible to provide an empirical assessment of their validity (such as the likelihood of a pattern of communications errors that could defeat the error control coding used). Assertions about the behaviour of both the low grade link (LGL) and baseband link (BBL) protocols are proved on the basis of these assumptions; this includes 'control behaviour', concerned with the control elements of the protocol, and 'information behaviour', concerned with the data carried and its integrity.

Other assertions stated but not proved would then be imposed as proof obligations of more detailed hardware and software design. For functional requirements, it was also necessary to make assertions about the 'well-behaved' nature of the environment (eg that the number of changes occurring could be supported by the information rate of the low grade link).

The proof required detailed aspects of system behaviour (including many aspects of timing, synchronisation and interference) to be addressed formally. It was only when developing the proof in this detail that the difficult design issues were really addressed fully. Processes on many timescales were modelled and assertions proved as part of the assurance process.

Study conclusions

The conclusions of this study have been reinforced by subsequent studies in which similar design problems have been addressed. The remainder of this paper attempts to draw together some of these wider conclusions in order to establish the way forward for these kinds of techniques.

Conclusions are in two broad areas: those regarding the processes of safety engineering and the ways in which formal techniques may help, and those regarding the nature of the formal techniques themselves.

The safety engineering process

We must focus on the development of proofs and not just specifications if formal methods are to underpin the safety engineering process fully.

In order to establish the effectiveness of any phase of development work, the engineer(s) responsible for the work must, at the appropriate review, explain firstly, how the various aspects of the design relate to the objectives set for it, and, secondly, that together, they satisfy them.

A formal approach assists this process in two ways. Firstly, a formal language provides a means of precise description. This allows the elements of the design, and the objectives themselves, to be documented unambiguously, so that the designers, reviewers and the implementors can all agree on the meaning of the design. Secondly, and more importantly, the mathematical foundation of the language means that, in principle, the argument justifying that the design objectives are satisfied can be provided in the form of a mathematical proof. This means that errors and inconsistencies in the argument can be identified and omissions, if not fully resolved, can be documented and subjected to further (possibly only empirical) analysis.

Emphasis is usually placed on the descriptive element of this process. Where formal methods are used to define protocol standards, for example, the emphasis is entirely on the descriptive accuracy of the formal language. Where formal methods are being used to support developments involving large teams, the accuracy of the description is potentially a clear benefit. However, the safety engineering process aims not only to produce a design, but also a safety argument. Every aspect of the development process is guided by the obligation to demonstrate the safety of the design. The most important product of the safety engineering process is the argument; the design is useless without it. In order to support this process, therefore, formal methods must contribute to the documentation and verification of the safety argument and capture it as a proof.

The use of fault-tree analysis, for example, formalises parts of the safety argument. A fault tree is a chain of assertions that proves that the top-level hazardous event cannot occur if particular combinations of underlying hazardous events are known not to be credible or possible. The production of a formal proof is motivated by the same needs. However, in the case of formal proof, the complexities of the events concerned, and their impact in combination, can be subject to more thorough and rigorous analysis.

Formal techniques must address issues of system safety and complexity of behaviour rather than those of software alone.

Safety engineering practice recognises the importance of system-level issues. Software alone cannot be a safety risk; it is only when software is placed in the context of a system, operating in a hazardous environment, that it becomes safety-critical. It is the complexity of this context that makes the task of design and verification so difficult. Software itself is not the problem; we only use software because it is the most effective way of building systems of the complexity that we require. The difficult design issues arise from this complexity, so it is these that must be overcome when assuring the safety of a system. Attempts to simplify the design process, or the systems employed, will not succeed because the complexity usually arises from the environment. Indeed, it is often the complexity of the environment that gives rise to the need for the system in the first place (as is the case with air traffic control systems, for example). The complexity of the relationship between the environment and the details of protocol timing described above is another example.

We should not try to hide the complexity by using some powerful mathematical technique; the best that can be achieved is to allow the complexity to be confronted and explored. Only then can verification be carried out with confidence.

Commensurate levels of assurance are required for each part of the development process. Therefore, whilst verification of code by static analysis, for example, meets a clear need, it does not address systems issues at all. Experience has shown that, particularly with real-time or reactive systems, that the difficult design issues do not arise in the software itself; the complexities of synchronisation and interference exist independently of software design and should be addressed, as system issues, by appropriate methods.

Design is creative rather than routine and methods must be formulated to underpin the engineering process, not substitute for it.

Production of the safety argument is the creative part of the design process. This is therefore the job of the engineer. Safety arguments, until explicitly documented, exist in the mind of the engineer and are developed on the basis of intuition and experience. No formal technique can, or should aim to, substitute for this creative process. Nor should the task of developing the argument be transferred from the engineer to the mathematician. Rather, a formal approach should be seen as underpinning the engineer's task. The mathematician should help the engineer to articulate his intuition so that it may be subjected to thorough analysis.

Researchers have for a long time attempted to formulate 'rules' or a method for producing specifications and designs. The options open to the designer at each stage would be constrained so as to guide him through the process. This, we believe, also runs counter to the principles of engineering practice. Instead of constraining the engineer, the existence of a formal underpinning should liberate him and allow him to develop novel and creative solutions with the confidence that, once the necessary verification of his argument has been carried out, a high level of safety assurance will be achieved.

A number of current approaches to software structuring for safety critical systems aim to simplify the software structure as much as possible, with the structure determined by the available verification techniques. Whilst simplification of software structure is to be encouraged, simplification of the safety argument should be a stronger guiding principle. Most current verification methods do not allow this to take place because they do not allow natural software and system structures to be adopted.

Supporting formal techniques

Formal techniques must be capable of abstract and modular description so that they can support the design process in a natural way.

Systems with complex behaviour are developed by assembling then from subsystems. The complexity of the interaction between subsystems is not something we should aim to eliminate; we need to exploit it because that is how complex systems are engineered. It follows that the properties of interest in the complete system cannot be traced to individual components, but arise from the way the various subsystems have been combined. It is not sensible to try to "separate concerns" because this is only half of the design process. We also need to address "combination of concerns". Examples arise in all branches of engineering. A mechanical engineer constructs a bridge from an assembly of smaller and simpler structures, wherein the integrity of the whole depends upon the ways in which the substructures are combined. The mechanical engineer simplifies the analysis of the bridge by considering the arrangement of substructures separately from the way each itself is constructed. The properties of interest emerge from the whole structure and cannot be attributed to any singe component. It is necessary to capture, in an abstract way, what each component *does* contribute to the structure, and then to establish the desired properties from the manner in which they have been combined.

The emergent properties of interest in a computer-based subsystem can often be more complex than the overall system safety properties. In order to support the analogue of the mechanical engineer's analysis of structures, these essential properties must be captured in an abstract way. These are often extremely difficult to formalise.

These considerations place significant demands upon the formalism used. The following qualities are required:

- it must be natural: it must be amenable to natural expression, by the engineer, of his intuition, in his own terms. In reality, the principles behind the proof cannot come from any other source.

- it must be faithful: the mathematical model underlying the language must be sufficiently faithful to the real world (eg an accurate model of the physics of the problem) so that maximum confidence can be placed on its conclusions. Impenetrable mathematical models bearing little relationship to the real world cannot inspire confidence.

- it must have expressive power commensurate with the problem domain: each problem will exhibit different aspects that need to be described. Concurrency and unpredictability of external systems need to be described if they impact the design of the system under development and formalisms must be capable of capturing the relevant properties.

- it must support abstraction: this is the only way in practice that complexity can be confronted and confined.

A multiplicity of views must be supported so that the necessary descriptive and analytical processes can be carried out.

The techniques we have been using distinguish between different 'views' of a system in which different types of assertion may be made. For example, views may be either:

- state-based or event-based: describing the state of the system or the events in which it participates;

- instantaneous or long term: an assertion may be made about a system that relates only to a specific instant in time, or may be made about the distribution of various events in time.

Assertions in each of the views need to be related to one another. For example, state-based and event-based views may be related using the idea of preconditions and postconditions for event specifications. Instantaneous and long-term views are related by the idea of historical inference: analysis of how long term behaviour is determined by the cumulative effect of many events over shorter timescales.

For example, the following assertions may be made about an oscillating dynamic system:

- $$\frac{d^2x}{dt^2} = -\omega^2(x-x_0);$$

- the behaviour of the variable x is periodic.

The first assertion is about the status of the system at an instant; the second is from a long term view relating behaviour at one time to that at other times. In this case, we know how to infer one assertion form the other; an analogous process of inference is required to develop safety arguments for systems where requirements are most naturally expressed in terms of long term behaviour.

Methods must include principles for structuring proofs, as it is the proofs that will govern the structure of the underlying specifications.

When developing safety critical systems, the obligation to provide the safety argument governs the structure of the system. When engineering is supported by formal analysis, the same must be true; the aim is to exploit the structure of the proof in structuring the system. The proof structures for the example project described above illustrate this. The problem of identifying 'proof components' alongside 'specification components' or 'implementation components' is currently only poorly understood and forms a major part of our research plans in order to facilitate exploitation of the techniques.

Summary

The proof is the main product of the safety engineering process. It is the objective of the analysis, not the specification or the design.

Safety is not simply about software. Software is used because it is the most effective way of building systems of the complexity that we need. Complexity is the difficulty that must be confronted, so methods must be designed from the point of view of addressing complexity per se, not software.

Safety is a system-level issue, and cannot be confined to analysis of software. Verification that software components meet design specifications does not verify safety. A commensurate level of assurance is required throughout the lifecycle. Therefore methods must be able to describe and analyse the complexity in the environment, and the interactions between the system and its environment, rather than just the systems themselves.

Proofs, ie safety arguments, come from the designer's intuition. Generation of the proof cannot be delegated to back-room mathematicians. The proof must be a natural expression of the argument, and methods must therefore be suited to natural expression.

The way ahead is to focus on tools and techniques to:

- develop proofs, not just specifications;

- address systems, not just software;

- confront complexity, not hide it;

- encourage creativity, not stifle it;

- underpin the engineering process, not substitute for it;

- and aim for the natural and not the impenetrable mathematical representation.

AUTHOR INDEX